Quantum cohomology
at the Mittag-Leffler Institute

a cura di P. Aluffi

APPUNTI

SCUOLA NORMALE SUPERIORE
1997

Preface

These are transcripts of notes taken at (some of) the lectures given at the Mittag-Leffler institute during the first semester of the 1996/97 year on *Enumerative geometry and its interaction with theoretical physics*. The first part of this collection consists of notes from talks on the basics of quantum cohomology, as developed in [F-P]. These talks formed the main body of the Tuesday seminar series at the Institute. The second part treats more advanced topics in quantum cohomology, which were primarily addressed in the Thursday seminar series. The third part consists of background material and related topics and contains material from both of these two series. An appendix, kindly provided by A. Kresch, gives a description of his C-program `farsta` for quantum cohomology computations.

These notes are meant as a series of snapshots of quantum cohomology as seen by the speakers at the time of their lectures. The reader should bear in mind that quantum cohomology is a growing and rapidly changing field; as any snapshot of a moving target, these notes are unavoidably a little blurry. Many of the writeups have been left in the form of the original talks, which were usually more concerned with giving motivations and a point of view, rather than conveying detailed proofs or attempting to survey the considerably extensive literature on the subject. Also, a glance at the references will show that many of the talks were based on preliminary (and hence not yet refereed) versions of papers on the subject. The published versions of these papers should be consulted for the definitive statements of the results, for the details of their proofs, and for more references.

Most of the notes were taken by Paolo Aluffi. Missing and complementary notes were contributed by P. Belorousski, C. Faber, B. Fantechi, W. Fulton, and S. di Rocco, all of whom are warmly thanked here. Many thanks are due to the speakers, both for preparing and delivering the talks, and also for glancing through a preliminary version of these notes and suggesting a number of corrections. Some of the speakers took the trouble of reworking the files for their talks themselves, and deserve a particular note of gratitude both for saving a great deal of work and for infinitely improving the final result.

We are very grateful to the organizers of the wonderfully successful 1996/7 year at the Mittag-Leffler institute, and to NSF for partial support (under grant #9500843) during the preparation of these notes.

A preliminary version of these notes appeared as Mittag-Leffler Report No. 10, 1996/7.

Paolo Aluffi

Contents

Part I

Stable Maps and Quantum Cohomology

1. Introduction—P. Belorousski
September 10, 1996

The aim is to go rather carefully through the definition of the quantum co-homology ring $QH^*(X)$ of a variety X (satisfying suitable hypotheses). This is a ring supported on $H^*(X) \otimes \mathbb{Q}[[y]]$, where y stands for a set of variables, and whose product is defined to reflect sophisticated enumerative information about X. The definition of this product relies on Gromov-Witten invariants, obtained by pulling back classes from X to suitable moduli spaces and intersecting them there. One of our main objectives will be to state precise assumptions and results about the existence and properties of these moduli spaces, postponing the proofs of these properties till after we have seen some enumerative applications.

We will be following [F-P] rather closely. The paper combines Fulton's notes on quantum cohomology [FultonSC] and Rahul Pandharipande's notes on the moduli spaces of stable maps; it is organized as follows:

0: Introduction;

1–6: Construction of the moduli spaces of maps; proof of the properties;

7: Gromov-Witten invariants;

8: Definition of quantum cohomology;

9,10: Applications to enumerative geometry and more.

We will go through 0, then jump to 7–10 extracting properties from 1–6 without proof, then go back to 1–6 and the proofs of those properties.

Background on moduli problems. One seeks to represent a contravariant functor $F :$ (Schemes) $\rightarrow$ (Sets) associating to each S the set $F(S)$ of equivalence classes of families of a certain kind of objects over S.

In the strongest possible sense, one would look for a *Fine moduli space*, that is a scheme X equipped with a universal family $U \rightarrow X$ and representing the functor. That is, "every" family of the prescribed objects would be obtained by pull-back of U from X.

This requirement is in general too strong. For example, the functor $\mathcal{M}_g$ for smooth genus-g curves does not have a fine moduli space.

The *Coarse moduli space* would be a scheme Y with a natural transformation from the functor F to the functor $h_Y = \mathrm{Hom}(-, Y)$, such that

(1) $F(\mathrm{Spec}\mathbb{C}) \cong h_Y(\mathbb{C})$ as sets; and
(2) Y satisfies the universal property of mapping uniquely to any other candidate Z (so that there exists a factorization $F \to h_Y \to h_Z$).

Coarse moduli spaces for interesting objects do exist: e.g., $\mathcal{M}_g$ has a coarse moduli space M_g (of dim. $3g - 3$ for $g > 1$).

We will deal with a whole hierarchy of moduli spaces, prescribing points on the curves ($M_{g,n}$); or parametrizing maps of curves to a given X and satisfying properties as detailed in the next lecture. We will construct and study good compactifications of these moduli spaces.

---

2. Stable maps—T. Graber
September 10, 1996

Defining the product in the quantum cohomology ring will require counting objects of the form $(C, p_1, \ldots, p_n, f)$ where C is a smooth curve of a given genus, p_i are prescribed smooth distinct points of C, and $f : C \to X$ is a map such that $f(p_i) \in$ prescribed loci, and $f_*[C] =$ prescribed $\beta \in A_1 X$. Quantum cohomology can be used for example to compute the number of degree-d rational curves in $X = \mathbb{P}^2$ which contain $3d - 1$ general points.

The general plan for these computations is the following:

(1) Construct a compactification of the appropriate moduli space;
(2) Do intersection theory on that space;
(3) Use this to define $QH^*(X)$ and solve enumerative problems.

DEFINITION. An n-pointed quasi-stable curve of genus g, $(C, p_1, \ldots, p_n)$, will be a projective, reduced, connected, at worst nodal curve C, with $h^1(\mathcal{O}_C) = g$, labeled with n marked distinct smooth points p_i.

A family of n-pointed quasi-stable curves over a scheme S will be a flat, projective map $\mathcal{C} \to S$ with n sections $p_1, \ldots, p_n : S \to \mathcal{C}$ such that each geometric fiber $(\mathcal{C}_s, p_1(s), \ldots, p_n(s))$ is quasi-stable. A family of maps to X will be a map $\mathcal{C} \to X$, and isomorphisms of families are defined in the obvious way respecting the data (and inducing the identity on the base space).

DEFINITION. A *stable* map of an n-pointed quasi-stable curve to X consists of the data $(C, p_1, \ldots, p_n, f)$ where $(C, p_1, \ldots, p_n)$ is as above, and $f : C \to X$ is such that

(1) all smooth irreducible rational components contracted to points in X have at least 3 'special' points; and
(2) all irreducible genus-1 components contracted to points have at least 1 special point.

Here 'special' means either marked, or on the intersection of the component with the closure of its complement. Condition (2) is nearly vacuous, in that it is a restriction only on 0-pointed, genus-1 curves which map to a point.

The definition is devised so that every $f : C \to X$ will have only finite automorphism group (where the notion of isomorphism of stable maps is defined in the obvious way).

We work over $\mathbb{C}$. The following theorems will be proved in these lectures:

THEOREM 1. *For any projective algebraic scheme X and every $\beta \in A_1 X$ there exists a coarse (compactified) projective moduli space, denoted $\overline{M}_{g,n}(X, \beta)$, parametrizing stable maps $(C, p_1, \dots, p_n, f)$ (with C of genus g) such that $f_*[C] = \beta$.*

For the moduli space to be reasonably well-behaved, some condition on X is required.

DEFINITION. A smooth projective variety X is called *convex* if for all maps $f : \mathbb{P}^1 \to X$, $H^1(f^*T_X) = 0$.

Important example: all homogeneous spaces are convex; so projective spaces, Grassmannians, flag manifolds, etc. are convex. (To see that homogeneous $\Longrightarrow$ convex: if X is homogeneous, then TX is generated by global sections, so f^*TX is generated by global sections in $\mathbb{P}^1$; writing $f^*TX = \oplus \mathcal{O}(d_i)$, we see all $d_i \geq 0$, and this implies that there is no H^1.)

THEOREM 2. *If X is a smooth projective convex variety, and $\beta \in A_1 X$, then $\overline{M}_{0,n}(X, \beta)$ is a variety of pure dimension*

$$\dim(X) + \int_\beta c_1(TX) + n - 3$$

(if nonempty), with at worst finite quotient singularities.

THEOREM 3. *The general element of $\overline{M}_{0,n}(X, \beta)$ does correspond to a map of an irreducible curve. The locus corresponding to reducible curves is a divisor with normal crossings (up to finite quotient singularities).*

Note: a priori, $\overline{M}_{0,n}(X, \beta)$ might be disconnected. The above statements hold for every irreducible component; these are necessarily disjoint. For X homogeneous, it can be shown that $\overline{M}_{0,n}(X, \beta)$ is in fact irreducible (Kim–Pandharipande, [Thomsen]).

EXAMPLES. — The simplest example in all X: $\beta = 0$. In this case, $\overline{M}_{g,n}(X, 0) = X \times \overline{M}_{g,n}$.

— If X is an abelian variety (so it contains no rational curves), then $\overline{M}_{0,n}(X, \beta) = \emptyset$ unless $\beta = 0$.

— $\overline{M}_{0,0}(\mathbb{P}^r, 1) = G(1, r)$, the Grassmannian of lines in $\mathbb{P}^r$. Indeed, there are no marked points, so no reducible curves can map stably in the above sense.

— $\overline{M}_{0,1}(\mathbb{P}^r, 1)$ = data of a line in $\mathbb{P}^r$, with a marked point. Again, no component may contract because there aren't enough marked points. So $\overline{M}_{0,1}(\mathbb{P}^r, 1)$ is the tautological line over $G(1, r)$.

— $\overline{M}_{0,0}(\mathbb{P}^2, 2)$ recovers the space of complete conics. Indeed, the general element is a smooth conic (up to automorphisms, a smooth conic has a unique parametrization); this can degenerate to a pair of lines, parametrized by a pair of intersecting lines in the obvious way; if the two components of the domain map to the same $\mathbb{P}^1$ in $\mathbb{P}^2$ we get a double line with a marked point (corresponding to the image of the intersection of the two components); and finally we find double lines with two marked points, arising from double covers (ramified at the two points).

---

3. Gromov–Witten invariants—J. Thomsen
September 24, 1996

We work over $\mathbb{C}$. For X a scheme, $\beta \in A_1 X$, and $g, n \geq 0$, we have the functor $\overline{\mathcal{M}}_{g,n}(X, \beta)$ from Algebraic Schemes to Sets, defined by

$$\overline{\mathcal{M}}_{g,n}(X, \beta)(S) = \left\{ \begin{array}{l} \text{stable families of maps over } S \\ \text{from } n\text{-pointed, genus-}g \text{ curves to } X, \\ \text{which represent } \beta \end{array} \right\}_{/\text{isomorphism}}$$

A stable family as above will be denoted $(\pi : \mathcal{C} \to S, \{p_i\}_{1 \leq i \leq n}, \mu : \mathcal{C} \to X)$; here

- (i) $\pi : \mathcal{C} \to S$ is flat, projective;
- (ii) $p_i : S \to \mathcal{C}$ are sections, $1 \leq i \leq n$;
- (iii) each geometric fiber of π, $\pi_s : \mathcal{C}_s \to \{s\}$ is a connected, nodal curve of arithmetic genus g and marked with distinct nonsingular points $p_1(s), \ldots, p_n(s)$;
- (iv) $\forall s \in S$, $\mu_s : \mathcal{C}_s \to X$ is stable: contracted genus-0 components have at least 3 'special' (marked or singular) points; contracted genus-1 components have at least 1 special point;
- (v) $\forall s \in S$, $\mu_{s_*}[\mathcal{C}_s] = \beta$.

From now on, X is taken to be projective. We will assume the following results:

THEOREM 1. *There exists a projective coarse moduli space $\overline{M}_{g,n}(X, \beta)$ for the functor $\overline{\mathcal{M}}_{g,n}(X, \beta)$.*

THEOREM 2. *Assume X is nonsingular and convex. Then $\overline{M}_{0,n}(X, \beta)$ is a fine moduli space for $\overline{\mathcal{M}}_{0,n}(X, \beta)$ away from curves with non-trivial automorphisms. More precisely:*

(i) *$\overline{M}_{0,n}(X, \beta)$ is a locally normal projective variety of pure dimension*

$$\dim X + \int_\beta c_1(T_X) + n - 3$$

(if nonempty);

(ii) $\overline{M}_{0,n}(X,\beta)$ *is locally a quotient of a nonsingular variety by a finite group;*

(iii) *The closed points in* $\overline{M}_{0,n}(X,\beta)$ *corresponding to irreducible curves form a dense open subscheme* $M_{0,n}(X,\beta)$ *of* $\overline{M}_{0,n}(X,\beta)$*;*

(iv) *There exists a nonsingular fine moduli space* $\overline{M}^*_{0,n}(X,\beta)$ *for automorphism- free curves;*

(v) $\overline{M}^*_{0,n}(X,\beta)$ *is an open subset of* $\overline{M}_{0,n}(X,\beta)$ *(dense if* $n \geq 3$*);*

(vi) *If* $\overline{M}_{0,n-1}(X,\beta) \neq \emptyset$*, then there exists a map* $\overline{M}_{0,n}(X,\beta) \xrightarrow{\pi} \overline{M}_{0,n-1}(X,\beta)$ *'forgetting' the first point. For s general in* $\overline{M}_{0,n-1}(X,\beta)$*, $C_s = \pi^{-1}(s)$ is the curve corresponding to s, and the composition $C_s \to \overline{M}_{0,n}(X,\beta) \xrightarrow{\rho_1} X$ is the corresponding map to X, where ρ_1 denotes the 'evaluation map' at the first point.*

The evaluation maps ρ_i, $1 \leq i \leq n$, are defined as follows. To give maps $\rho_i : \overline{M}_{0,n}(X,\beta) \to X$ amounts to giving natural transformations

$$\theta_i : \overline{\mathcal{M}}_{0,n}(X,\beta) \to \mathrm{Hom}(*,X) \quad :$$

at a set S, $\theta_i(S)$ sends a stable family over S:

$$\mathcal{C} \xrightarrow{\mu} X$$
$$\downarrow$$
$$S$$

with sections $p_1, \ldots, p_n$, to the composition

$$S \xrightarrow{p_i} \mathcal{C} \xrightarrow{\mu} X$$

For $S = \mathrm{Spec}\,\mathbb{C}$, that is for a single stable curve C, this simply evaluates the map μ to X at $p_i \in C$.

For a permutation $\sigma \in S_n$, there is an automorphism of $\overline{M}_{0,n}(X,\beta)$ interchanging the points:

$$\overline{M}_{0,n}(X,\beta) \xrightarrow{\Sigma(\sigma)} \overline{M}_{0,n}(X,\beta)$$

This is induced by the natural transformation sending a family

$$\mathcal{C} \xrightarrow{\mu} X$$
$$\downarrow$$
$$S$$

with sections $p_1, \ldots, p_n$, to the same diagram but with sections $p_{\sigma(1)}, \ldots, p_{\sigma(n)}$. It is clear that

$$\rho_{\sigma(i)} = \rho_i \circ \Sigma(\sigma)$$

From now on, assume X is projective, nonsingular and convex.

DEFINITION. (Gromov-Witten invariants.) Given $\beta \in A_1 X$, and $\gamma_1, \ldots, \gamma_n \in A^* X$ (the *operational* Chow ring; $\int_A$ will denote evaluation at $A \in A_*$), define

$$I_\beta(\gamma_1 \cdots \gamma_n) = \int_{\overline{M}_{0,n}(X,\beta)} \rho_1^*(\gamma_1) \cup \cdots \cup \rho_n^*(\gamma_n)$$

REMARKS.

(1) If the γ_i are homogeneous, $I_\beta(\gamma_1 \cdots \gamma_n) = 0$ if $\sum_i \operatorname{codim} \gamma_i \neq \dim \overline{M}_{0,n}(X,\beta)$;

(2) Due to the presence of the S_n-action on $\overline{M}_{0,n}(X,\beta)$, $I_\beta(\gamma_1 \cdots \gamma_n)$ is invariant under permutation of the γ_i's.

LEMMA. *Assume X is homogeneous: $X = G/P$; and $\Gamma_1, \ldots, \Gamma_n \subset X$ ($n \geq 3$) are subvarieties, such that $\sum \operatorname{codim}(\Gamma_i) = \dim \overline{M}_{0,n}(X,\beta)$. Then for a general $\sigma = (g_1, \ldots, g_n) \in G^n$, the scheme-theoretic intersection*

$$\rho_1^{-1}(g_1 \Gamma_1) \cap \cdots \cap \rho_n^{-1}(g_n \Gamma_n)$$

consists of reduced points supported on the locus $M_{0,n}(X,\beta)$. Further, the number of points in this intersection equals $I_\beta(\check{\Gamma}_1 \cdots \check{\Gamma}_n)$.

PROOF. This follows from judicious use of Kleiman-Bertini [Kleiman]. Let G be a connected algebraic group, and X a homogeneous G-variety, Y, Z varieties mapping to X:

$$Z$$
$$\downarrow g$$
$$Y \xrightarrow{\ f\ } X$$

For $\sigma \in G$, denote by Y^σ the image of the composition $Y \xrightarrow{f} X \xrightarrow{\sigma} X$. Then

(1) There is a dense open subscheme G° of G such that, for $\sigma \in G^\circ$, $Y^\sigma \times_X Z$ is either empty, or of pure dimension

$$\dim Y + \dim Z - \dim X$$

(2) Further, if Y and Z are nonsingular, then G° can be found so that $Y^\sigma \times_X Z$ is nonsingular.

This is used four times in our proof of the Lemma. Notations: $\Gamma = \Gamma_1 \times \cdots \times \Gamma_n$; $\Gamma^\circ \subset \Gamma$ is the nonsingular locus; Γ_{sing} is the singular locus of Γ; $M_{0,n}(X,\beta)^c$ is the complement of $M_{0,n}(X,\beta)$ in $\overline{M}_{0,n}(X,\beta)$. Four applications of [Kleiman]:

$$\text{(I)} \qquad \overline{M}_{0,n}(X,\beta)$$
$$\downarrow (\rho_1,\ldots,\rho_n) \quad :$$
$$\Gamma \longrightarrow X^n$$

for a general $\sigma \in G^n$, $\Gamma^\sigma \times_{X^n} \overline{M}_{0,n}(X,\beta)$ is either empty, or of dimension 0;

$$(\text{II}) \qquad \overline{M}_{0,n}(X,\beta)$$
$$\downarrow \qquad :$$
$$\Gamma_{\text{sing}} \longrightarrow \qquad X^n$$

for a general $\sigma \in G^n$, $\Gamma^\sigma_{\text{sing}} \times_{X^n} \overline{M}_{0,n}(X,\beta)$ is either empty, or of dimension < 0; hence, it is necessarily empty. That is, we may assume that $\Gamma = \Gamma^\circ$ is nonsingular;

$$(\text{III}) \qquad M_{0,n}(X,\beta)^c$$
$$\downarrow \qquad :$$
$$\Gamma \longrightarrow \qquad X^n$$

for a general $\sigma \in G^n$, $\Gamma^\sigma \times_{X^n} M_{0,n}(X,\beta)^c$ is again empty, or of dimension < 0; hence, empty. So the intersection is supported on $M_{0,n}(X,\beta)$. Finally

$$(\text{IV}) \qquad M_{0,n}(X,\beta)$$
$$\downarrow \qquad :$$
$$\Gamma^\circ \longrightarrow \qquad X^n$$

for a general $\sigma \in G^n$, $\Gamma^{\circ \sigma} \times_{X^n} M_{0,n}(X,\beta)$ is smooth, hence reduced, of dimension 0 (if nonempty).

Summarizing we have that, for a general $\sigma = (g_1, \ldots, g_n) \in G$,

$$\rho_1^{-1}(g_1\Gamma_1) \cap \cdots \cap \rho_n^{-1}(g_n\Gamma_n) = \Gamma^\sigma \times_{X^n} \overline{M}_{0,n}(X,\beta)$$

is reduced, of dimension 0, and supported on the nonsingular subset $M_{0,n}(X,\beta)$. To see that the number of points in this intersection equals $I_\beta(\check{\Gamma}_1 \cdots \check{\Gamma}_n)$, observe that we have the fiber square

$$\rho_1^{-1}(g_1\Gamma_1) \cap \cdots \cap \rho_n^{-1}(g_n\Gamma_n) \longrightarrow \overline{M}_{0,n}(X,\beta) \times \Gamma^\sigma$$
$$\downarrow \qquad\qquad\qquad\qquad\qquad\qquad \downarrow$$
$$\overline{M}_{0,n}(X,\beta) \xrightarrow{\;i=(id,\rho)\;} \overline{M}_{0,n}(X,\beta) \times X^n$$

with the bottom line a regular embedding, of codimension $n \dim X$. As the dimensions match, and as $\rho_1^{-1}(g_1\Gamma_1) \cap \cdots \cap \rho_n^{-1}(g_n\Gamma_n)$ is reduced,

$$[\rho_1^{-1}(g_1\Gamma_1) \cap \cdots \cap \rho_n^{-1}(g_n\Gamma_n)] = i^!(\overline{M}_{0,n}(X,\beta) \times \Gamma^\sigma)$$

where $i^!$ denotes the gysin map. On the other hand, $i^!(\overline{M}_{0,n}(X,\beta) \times \Gamma^\sigma) = \psi(g_1\Gamma_1 \times \cdots \times g_n\Gamma_n)$, where ψ is the composition

$$A_*(X)^{\otimes n} \longrightarrow A_*(X^n) \xrightarrow{\;\cong\;} A^*(X^n) \xrightarrow{\;\rho^*\;} A^*(\overline{M}_{0,n}(X,\beta)) \xrightarrow{\;\cap[\overline{M}_{0,n}(X,\beta)]\;} A_*(\overline{M}_{0,n}(X,\beta))$$

Tracing the definition, $\psi(g_1\Gamma_1 \times \cdots \times g_n\Gamma_n)$ equals $I_\beta((g_1\Gamma_1)^\vee \cdots (g_n\Gamma_n)^\vee)$; finally, $[\Gamma_i] = [g_i\Gamma_i]$ and the result follows. $\square$

The Gromov-Witten invariants satisfy a number of general properties. For example: assume X is homogeneous, and $\gamma_1 \in A^1(X)$ and $n \geq 4$ (or $n \geq 1$ if $\beta \neq 0$). Then

$$I_\beta(\gamma_1 \cdots \gamma_n) = \left(\int_\beta \gamma_1\right) I_\beta(\gamma_2 \cdots \gamma_n)$$

Indeed, consider $\phi : \overline{M}_{0,n}(X,\beta) \xrightarrow{\rho_1 \times \text{forget.}} X \times \overline{M}_{0,n-1}(X,\beta)$; then write

$$\phi_*[\overline{M}_{0,n}(X,\beta)] = \beta' \times [\overline{M}_{0,n-1}(X,\beta)] + \alpha$$

where α dominates a proper subset of $\overline{M}_{0,n-1}(X,\beta)$. (Here we use that X is homogeneous, so that $A_*(X \times \overline{M}_{0,n-1}(X,\beta)) = A_*(X) \otimes A_*(\overline{M}_{0,n-1}(X,\beta))$.) Pushing forward doesn't change degrees, so

$$I_\beta(\gamma_1 \cdots \gamma_n) = \int_{[\overline{M}_{0,n}(X,\beta)]} \rho_1^*(\gamma_1) \cup \cdots \cup \rho_n^*(\gamma_n)$$

$$= \int_{\phi_*[\overline{M}_{0,n}(X,\beta)]} \gamma_1 \times (\rho_2^*(\gamma_2) \cup \cdots \cup \rho_n^*(\gamma_n))$$

$$= \int_{\beta' \times [\overline{M}_{0,n-1}(X,\beta)] + \alpha} \gamma_1 \times (\rho_2^*(\gamma_2) \cup \cdots \cup \rho_n^*(\gamma_n))$$

$$= \left(\int_{\beta'} \gamma_1\right) \int_{[\overline{M}_{0,n-1}(X,\beta)]} \rho_2^*(\gamma_2) \cup \cdots \cup \rho_n^*(\gamma_n)$$

by the projection formula: α is killed after push-forward to $\overline{M}_{0,n-1}(X,\beta)$

$$= \left(\int_{\beta'} \gamma_1\right) I_\beta(\gamma_2 \cdots \gamma_n)$$

Now the claim is that $\beta' = \beta$. To see this, consider a generic point $pt = (C, p_2, \ldots, p_n, \mu)$ of $\overline{M}_{0,n-1}(X,\beta)$. Using Theorem 2(vi) we get the fiber diagram

$$\begin{array}{ccc}
C & \longrightarrow & \overline{M}_{0,n}(X,\beta) \\
{\scriptstyle (\mu,pt)}\downarrow & & \downarrow{\scriptstyle \phi} \\
X \times \{pt\} & \longrightarrow & X \times \overline{M}_{0,n-1}(X,\beta) \\
\downarrow & & \downarrow{\scriptstyle p_2} \\
\{pt\} & \xrightarrow{\ j\ } & \overline{M}_{0,n-1}(X,\beta)
\end{array}$$

For a generic pt, j is a regular embedding. Therefore

$$j^!\phi_*[\overline{M}_{0,n}(X,\beta)] = (\mu,pt)_*j^![\overline{M}_{0,n}(X,\beta)] = (\mu,pt)_*[C] = \beta \times \{pt\}$$

on the other hand,

$$j^! \phi_* [\overline{M}_{0,n}(X,\beta)] = j^!(\beta' \times [\overline{M}_{0,n-1}(X,\beta)] + \alpha)$$
$$= j!(\beta' \times [\overline{M}_{0,n-1}(X,\beta)]) = \beta' \times \{pt\}$$

(no contribution from α, since it will miss a general pt). Hence $\beta \times \{pt\} = \beta' \times \{pt\}$, as needed.

---

4. Associativity of the quantum product—K. Ranestad
October 1, 1996

Usual notations: X is a variety, $\beta \in A_1(X)$, $\gamma_i \in A^*(X)$, μ is a map from $\mathbb{P}^1$ to X, or from a tree of $\mathbb{P}^1$'s; $p_1, \ldots, p_n$ are marked points, $\rho_i(*) = \mu(p_i)$ for $* \in \overline{M}_{0,n}(X,\beta)$.

Conditions on X:

(1) X is smooth, projective, convex (that is: $\forall \mu$, $h^1(\mathbb{P}^1, \mu^* T_X) = 0$);

(2) the Chow and topological homology theories of X are isomorphic;

(3) the effective cone in $A_1(X)$ is $\{\sum_{i=1}^p a_i \beta_i : a_i \in \mathbb{Z}_{\geq 0}, \beta_i$ in the form $\mu_*[\mathbb{P}^1]\}$.

The Gromov-Witten invariants are the intersection numbers

$$I_\beta(\gamma_1 \cdots \gamma_n) = \int_{\overline{M}_{0,n}(X,\beta)} \rho_1^*(\gamma_1) \cup \cdots \cup \rho_n^*(\gamma_n) \qquad \in \mathbb{Z}$$

We will need two properties of these numbers; a third one was stated and proved in Thomsen's lecture:

(1) if $\beta = 0$, so $\overline{M}_{0,n}(X,0) = \overline{M}_{0,n} \times X$, then $I_0(\gamma_1 \gamma_2 \gamma_3) = \int_X \gamma_1 \cup \gamma_2 \cup \gamma_3$, and $I_0(\gamma_1 \cdots \gamma_n) = 0$ for $n > 3$;

(2) if $\gamma_1 = 1$: $I_\beta(1, \gamma_2, \ldots, \gamma_n) = \begin{cases} 0 & n > 3 \text{ or } \beta \neq 0 \\ \int_X \gamma_2 \cup \gamma_3 \text{ otherwise} \end{cases}$.

Both of these follow easily from the projection formula.

We can write the classical intersection product in a fancy way, using these notations. Pick generators $T_0 = 1, T_1, \ldots, T_m$ for $A^* X$, set $g_{ij} = \int_X T_i \cup T_j$, and let g^{ij} be the inverse matrix of g_{ij} (the inverse exists by Poincaré duality, and since by assumption (2) we have no torsion). Via the isomorphism $A^*(X \times X) = A^* X \otimes A^* X$, the class of the diagonal $\Delta \subset X \times X$ is

$$[\Delta] = \sum g^{ij} T_i \otimes T_j$$

Denoting by p_1, p_2 the two projections, we have then

$$T_i \cup T_j = p_{2*}(p_1^*(T_i \cup T_j) \cup \Delta)$$

$$= p_{2*}\left(\sum_{e,f}(T_i \cup T_j \cup T_e) \otimes g^{ef}T_f\right)$$

$$= \sum_{e,f}\left(\int_X T_i \cup T_j \cup T_e\right) g^{ef}T_f$$

$$= \sum_{e,f} I_0(T_i T_j T_e) g^{ef}T_f$$

With this understood, set for $\gamma \in A^*X$:

$$\phi(\gamma) = \sum_{n \geq 3} \frac{1}{n!} \sum_\beta I_\beta(\gamma^n)$$

Writing $\gamma = \sum_{i=0}^m y_i T_i$, this expands to

$$\phi(\gamma) = \sum_{n_0 + \cdots + n_m \geq 3} \sum_\beta I_\beta(T_0^{n_0} \cdots T_m^{n_m}) \frac{y_0^{n_0}}{n_0!} \cdots \frac{y_m^{n_m}}{n_m!}$$

LEMMA. *This is a power series. In fact, for given n, $I_\beta(\gamma^n)$ is nonzero only for finitely many β.*

PROOF. This uses assumptions (1) and (3) on X. Since X is convex, we have $h^1(\mu^*T_X) = 0$ for all maps μ from $\mathbb{P}^1$ to X. Writing $\mu^*T_X = \oplus_{i=1}^{\dim X}\mathcal{O}(d_i)$ and perhaps composing first μ with a cover of $\mathbb{P}^1$, we may assume all $d_i \geq 0$. Also, for $\mu_*[\mathbb{P}^1] \neq 0$, the differential $d\mu : T_{\mathbb{P}^1} \to \mu^*T_X$ is injective, so at least one d_i is ≥ 2. Representing an effective β as a nonnegative combination of push-forwards of $\mathbb{P}^1$'s, we see that $\int_\beta c_1(T_X) \geq 2$. Also, as the effective cone is finitely generated we see that for a given N there are only a finite number of effective β for which $\int_\beta c_1(T_X) \leq N$. But if $I_\beta(\gamma^n)$ is nonzero, then $\dim X + \int_\beta c_1(T_X) + n - 3 = \dim \overline{M}_{0,n}(X, \beta) \leq n \cdot \mathrm{codim}\,\gamma \leq n \dim X$, and this bounds $\int_\beta c_1(T_X)$. □

Now we define the quantum product $*$. Consider the third partial derivatives

$$\phi_{ijk} := \frac{\partial^3 \phi}{\partial y_i \partial y_j \partial y_k}$$

$$= \sum_{n_0 + \cdots + n_m \geq 0} \sum_\beta I_\beta(T_0^{n_0} \cdots T_m^{n_m} T_i T_j T_k) \frac{y_0^{n_0} \cdots y_m^{n_m}}{n_0! \cdots n_m!}$$

$$= \sum_n \frac{1}{n!} \sum_\beta I_\beta(\gamma^n T_i T_j T_k) \quad ,$$

a power series in $y_0, \ldots, y_m$; we set

$$T_i * T_j := \sum_{e,f} \phi_{ije} g^{ef} T_f \qquad \in A^* X \otimes \mathbb{Q}[[y_0, \ldots, y_m]]$$

This product is clearly commutative, and has $T_0 = 1$ as unit by property (2) of the invariants:

$$T_0 * T_j = \sum_{e,f} \phi_{0je} g^{ef} T_f = \sum_{e,f} \sum_{n \geq 0} \frac{1}{n!} \sum_{\beta} I_{\beta}(\gamma^n 1 T_j T_e) g^{ef} T_f$$

$$= \sum_{e,f} I_0(1 T_j T_e) g^{ef} T_f = \sum_{ef} \left(\int_X T_j \cup T_e \right) g^{ef} T_f$$

$$= \sum_{e,f} g_{je} g^{ef} T_f = T_j$$

Associativity of $*$. The quantum product defined above is associative. To check this, first write out what it means:

$$(T_i * T_j) * T_k = \left(\sum_{e,f} \phi_{ije} g^{ef} T_f \right) * T_k = \sum_{e,f,c,d} \phi_{ije} g^{ef} \phi_{fkc} g^{cd} T_d$$

$$T_i * (T_j * T_k) = T_i * \left(\sum_{e,f} \phi_{jke} g^{ef} T_f \right) = \sum_{e,f,c,d} \phi_{jke} g^{ef} \phi_{ifc} g^{cd} T_d$$

That is, setting

$$F(ij|k\ell) = \sum_{e,f} \phi_{ije} g^{ef} \phi_{fk\ell} \quad ,$$

we need to show that

$$F(ij|k\ell) = F(jk|i\ell) \qquad \forall \ell$$

Expanding:

$$F(ij|k\ell) = \sum_{\beta_1, \beta_2; n_i \geq 0; 0 \leq e, f \leq m} \frac{1}{n_1! n_2!} I_{\beta_1}(\gamma^{n_1} T_i T_j T_e) g^{ef} I_{\beta_2}(\gamma^{n_2} T_f T_k T_\ell)$$

Now, the boundary of $\overline{M} = \overline{M}_{0,n}(X, \beta)$ consists of irreducible divisors $D = D(A, B, \beta_1, \beta_2)$, where $A \amalg B = \{1, \ldots, n\}$, and $\beta_1 + \beta_2 = \beta$. The general point of this divisor has a description similar to the case of $\overline{M}_{0,n}$ (see §3), by glueing stable curves at a point. More precisely,

$$D \cong \overline{M}_{0, A \cup \{\bullet\}}(X, \beta_1) \times_X \overline{M}_{0, B \cup \{\bullet\}}(X, \beta_2)$$

when $A \neq \emptyset$, $B \neq \emptyset$, with obvious notations. Setting $\overline{M}_A = \overline{M}_{0,A\cup\{\bullet\}}(X,\beta_1)$, etc. we have the diagram

$$
\begin{array}{ccccc}
\overline{M} & \xleftarrow{\;\alpha\;} & D & \xrightarrow{\;i\;} & \overline{M}_A \times_C \overline{M}_B \\
\rho\downarrow & & \varphi\downarrow & & \downarrow\rho' \\
X^n & \xleftarrow{\;\alpha'\;} & X^{n+1} & \xrightarrow{\;i'\;} & X^{n+2}
\end{array}
$$

with the right square a fiber square. Chasing this diagram gives:

LEMMA.

$$
i_*\alpha^*(\rho_1^*(\gamma_1)\cup\cdots\cup\rho_n^*(\gamma_n)) = \sum_{e,f} g^{ef}\left(\prod_{a\in A}\rho_a^*(\gamma_a)\rho_\bullet^*(T_e)\right)\left(\prod_{b\in B}\rho_b^*(\gamma_b)\rho_\bullet^*(T_f)\right)
$$

Now fix $\beta \in A_1 X$, $\gamma_1,\ldots,\gamma_n \in A^*X$, integers q,r,s,t in $\{1,\ldots,n\}$, and put

$$
G(qr|st) = \sum_{e,f;\beta_1+\beta_2=\beta;A\amalg B=\{1,\ldots,n\};q,r\in A;s,t\in B} I_{\beta_1}\left(\prod_{a\in A}\gamma_a T_e\right)g^{ef}I_{\beta_2}\left(\prod_{b\in B}\gamma_b T_f\right)
$$

By the Lemma,

$$
G(qr|st) = \sum\int_{D(A,B;\beta_1,\beta_2)}\rho_1^*(\gamma_1)\cup\cdots\cup\rho_n^*(\gamma_n) = \int_{D(qr|st)}\rho_1^*(\gamma_1)\cup\cdots\cup\rho_n^*(\gamma_n)
$$

where we denote by $D(qr|st)$ the sum of the relevant boundary divisors.

Now for the key observation: denoting by π the composition of the forgetful maps

$$
\overline{M}_{0,n}(X,\beta) \to \overline{M}_{0,n} \to \overline{M}_{0,\{q,r,s,t\}} \cong \mathbb{P}^1
$$

one checks that $D(qr|st)$ is the inverse image via π of the divisor of $\overline{M}_{0,\{q,r,s,t\}}$ corresponding to the partition $\{q,r\} \cup \{s,t\}$ (the convexity of X is used to show that all components in the preimage appear with multiplicity one). In particular,

$$
D(qr|st) \cong D(rs|qt)
$$

up to linear equivalence, since this equality holds in $\overline{M}_{0,\{q,r,s,t\}} \cong \mathbb{P}^1$; and therefore

$$
G(qr|st) = G(rs|qt)
$$

for all $\gamma_1,\ldots,\gamma_n$. Setting all but $\gamma_q, \gamma_r, \gamma_s, \gamma_t$ equal to γ, we have by definition

$$
G(qr|st) = \sum_{e,f;n_1+n_2=n;n_i\geq 2}\binom{n-4}{n_1-2}I_{\beta_1}(\gamma^{n_1-2}\gamma_q\gamma_r T_e)g^{ef}I_{\beta_2}(\gamma^{n_2-2}\gamma_s\gamma_t T_f)
$$

or, by a shift of the indices:

$$
G(qr|st) = (n-4)!\sum_{e,f;n_1+n_2=n-4;n_i\geq 0}\frac{1}{n_1!n_2!}I_{\beta_1}(\gamma^{n_1}\gamma_q\gamma_r T_e)g^{ef}I_{\beta_2}(\gamma^{n_2}\gamma_s\gamma_t T_f)
$$

Therefore, the equality of G's says

$$\sum_{e,f;n_1+n_2=n-4;n_i\geq 0} \frac{1}{n_1!n_2!} I_{\beta_1}(\gamma^{n_1}\gamma_q\gamma_r T_e)g^{ef}I_{\beta_2}(\gamma^{n_2}\gamma_s\gamma_t T_f)$$

$$= \sum_{e,f;n_1+n_2=n-4;n_i\geq 0} \frac{1}{n_1!n_2!} I_{\beta_1}(\gamma^{n_1}\gamma_r\gamma_s T_e)g^{ef}I_{\beta_2}(\gamma^{n_2}\gamma_q\gamma_t T_f) \quad ;$$

setting $\gamma_q = T_i$, $\gamma_r = T_j$, $\gamma_s = T_k$, $\gamma_t = T_\ell$ and adding over n:

$$\sum_{e,f;n_1,n_2\geq 0} \frac{1}{n_1!n_2!} I_{\beta_1}(\gamma^{n_1}T_iT_jT_e)g^{ef}I_{\beta_2}(\gamma^{n_2}T_kT_\ell T_f)$$

$$= \sum_{e,f;n_1,n_2\geq 0} \frac{1}{n_1!n_2!} I_{\beta_1}(\gamma^{n_1}T_jT_kT_e)g^{ef}I_{\beta_2}(\gamma^{n_2}T_iT_\ell T_f) \quad ;$$

and finally, adding over all $\beta_1 + \beta_2 = \beta$:

$$F(ij|k\ell) = F(jk|i\ell)$$

as needed.   □

---

5. Applications of QH^* to enumerative geometry—S.L. Kleiman
October 8, 1996

Claude Itzykson, in memoriam

The main reference for the following material is [F-P], especially §9; a secondary reference is [DF-I]. The material is organized into these sections:

—§1. *Gromov–Witten invariants*
—§2. *The potential*
—§3. *The projective plane*
—§4. *Feynman diagrams*
—§5. *Surfaces in general*
—§6. *Del Pezzo and Hirzebruch surfaces*

§1. Gromov–Witten invariants. Let X be a smooth irreducible projective variety. If X is, in fact, the quotient G/P of a reductive group G by a parabolic subgroup P, then X is convex and its singular homology $H_*(X, \mathbb{Z})$ is algebraic. As we have seen, it follows that there are good moduli spaces $\overline{M} = \overline{M}_{0,n}(X, \beta)$ of marked Kontsevich-stable maps, and that X has an associative quantum cohomology ring $QH^*(X)$. Later (in §5) we'll see that conversely, at least for surfaces, if X is convex and $H_*(X, \mathbb{Z})$ is algebraic, then, for all practical purposes, $X = G/P$.

Let $\beta \in A_1 X := H_2(X, \mathbb{Z})$. The moduli space $\overline{M}$ parameterizes maps of class β:

$$\overline{M} := \left\{ (C \xrightarrow{\mu} X; p_1, \ldots, p_n) \left| \begin{array}{l} C = \mathbb{P}^1 \text{ or a tree of } \mathbb{P}^1\text{s,} \\ \mu_*[C] = \beta, \; p_i \in C, \\ \text{and } \mu \text{ is Kontsevich-stable} \end{array} \right. \right\} \Big/ \text{ isom}$$

The *expected dimension* of $\overline{M}$ is

$$\exp. \dim(\overline{M}) := \dim X + \int_\beta c_1(TX) + n - 3.$$

Moreover, there are *evaluation maps* $\rho_i \colon \overline{M} \to X$, which take an element of $\overline{M}$ as above to $\mu(p_i)$. Note that, if $\overline{M}$ is nonempty, then β is *effective*; that is, β is the class of the image of a tree C of $\mathbb{P}^1$s.

Let $\gamma_1, \ldots, \gamma_n \in A^* X := H^*(X, \mathbb{Z})$. The corresponding *Gromov–Witten invariant* is

$$I_\beta(\gamma_1 \cdots \gamma_n) := \begin{cases} 0, & \text{unless } \beta \text{ is effective and} \\ & \sum \dim \gamma_i = \exp. \dim(\overline{M}); \\ \int_\varphi \rho_1^* \gamma_1 \cup \cdots \cup \rho_n^* \gamma_n, & \text{if so.} \end{cases}$$

Here φ is the fundamental class of the moduli space: $\varphi := [\overline{M}]$ if X is convex; φ is the 'virtual fundamental class' otherwise [Li-Tian], [B-F] and [Behrend] (see §6 for some examples). This class may even be fractional.

At least in the case $X = G/P$, we have

$$I_\beta(\gamma_1 \cdots \gamma_n) = \#(\rho_1^{-1}\Gamma_1 \cap \cdots \cap \rho_n^{-1}\Gamma_n),$$

where the Γ_i are representatives, in general position, of the Poincaré duals to the γ_i. In other words, I_β is the number of pointed maps μ such that $\mu_*[C] = \beta$ and $\mu p_i \in \Gamma_i$. Moreover, the intersection on the right consists of general points of $\overline{M}$. In particular, the corresponding C are each equal to $\mathbb{P}^1$. Now, it may be that the general map μ is not birational onto its image. For example, this is the case if $X = \mathbb{P}^1$ and $\beta = d[X]$ with $d \geq 2$, or if $X = \mathbb{P}^1 \times \mathbb{P}^1$ and $\beta = d[\text{ruling}]$ with $d \geq 2$. However, if the general map is not birational, then the intersection would not be finite unless it's empty; hence, it's empty, and $I_\beta(\gamma_1 \cdots \gamma_n)$ vanishes. Thus I_β is the number of irreducible rational curves in $X \; (= G/P)$ of class β meeting the Γ_i.

In general, the Gromov–Witten invariants possess the following three properties:

(I) $$I_0(\gamma_1 \cdots \gamma_n) = \begin{cases} 0, & \text{if } n > 3; \\ \int_X \gamma_1 \cup \gamma_2 \cup \gamma_3, & \text{if } n = 3. \end{cases}$$

(II) $$I_\beta(1 \cdot \gamma_2 \cdots \gamma_n) = \begin{cases} 0, & \text{if } n > 3 \text{ or } \beta \neq 0; \\ \int_X \gamma_2 \cup \gamma_3, & \text{if } n = 3 \text{ and } \beta = 0. \end{cases}$$

If γ_1 is a divisor class ($\gamma_1 \in A^1 X$), then

(III) $$I_\beta(\gamma_1 \gamma_2 \cdots \gamma_n) = \left(\int_\beta \gamma_1 \right) I_\beta(\gamma_2 \cdots \gamma_n).$$

§2. The potential.

The *potential* is defined by the formula,

$$\Phi(\gamma) = \sum_{n \geq 3} \sum_{\beta} \frac{1}{n!} I_{\beta}(\gamma^n).$$

Take a graded basis of $A^* X$ with $T_0 = 1$, with $T_1, \ldots, T_p \in A^1 X$, and with $T_{p+1}, \ldots, T_m$ of higher degree. Write $\gamma = \sum y_i T_i$, and use the linearity of I_{β} to expand Φ:

$$\Phi(y_0, \ldots, y_m) = \sum_{n_0 + \cdots + n_m \geq 3} \sum_{\beta} I_{\beta}(T_0^{n_0} \cdots T_m^{n_m}) \frac{y_0^{n_0}}{n_0!} \cdots \frac{y_m^{n_m}}{n_m!}.$$

The right-hand side is a well-defined formal power series in $\mathbb{Q}[[y]]$ if, given γ^n, we have $I_{\beta}(\gamma^n) = 0$ for almost all β. For example, this is the case if the effective classes lie in the cone generated by finitely many β such that $\int_{\beta} c_1 TX > 0$ (which is the key ingredient in the first lemma in the preceding lecture of Ranestad's). This condition is satisfied if $X = G/P$, or more generally whenever $-K_X := c_1 TX$ is ample. However, this issue becomes unimportant when Φ is modified as described next (see [G-P, §2]).

Break up the potential into two pieces, $\Phi = \Phi^{cl} + \Phi^{qtm}$, where Φ^{cl}, the *classical part*, is the contribution due to the condition $\beta = 0$, and where Φ^{qtm}, the *quantum correction*, is the contribution coming from all nonzero β. Property (I) of §1 yields

$$\Phi^{cl} = \frac{1}{3!} \int_X \gamma^3 = \sum_{n_0 + \cdots + n_m = 3} \int_X T_0^{n_0} \cup \cdots \cup T_m^{n_m} \frac{y_0^{n_0}}{n_0!} \cdots \frac{y_m^{n_m}}{n_m!}.$$

Since the quantum product $*$ (recalled below) involves only the third derivatives of Φ, we may modify Φ by terms of degree at most two. So (I)–(III) imply that we may replace Φ^{qtm} by

$$\Gamma := \sum_{n_{p+1} + \cdots + n_m \geq 3} \sum_{\beta \neq 0} I_{\beta}(T_{p+1}^{n_{p+1}} \cdots T_m^{n_m}) \left(\prod_{i=1}^{p} e^{y_i \int_{\beta} T_i} \right) \frac{y_{p+1}^{n_{p+1}}}{n_{p+1}!} \cdots \frac{y_m^{n_m}}{n_m!},$$

which is a formal power series in $y_{p+1}, \ldots, y_m$ and in new variables $e^{y_1}, \ldots, e^{y_p}$ (with appropriate derivatives) and in their inverses. The product $\prod_{i=1}^{p} e^{\int_{\beta} y_i T_i}$ may be abbreviated to $e^{\int_{\beta} \gamma}$ as only the T_i in $A^1 X$ give nonzero contributions to the integral.

Set $\Phi_{ijk} := \dfrac{\partial^3 \Phi}{\partial y_i \partial y_j \partial y_k}$, set $(g^{ef}) := (g_{ef})^{-1}$ for $g_{ef} = \int_X T_e \cup T_f$, and set

$$T_i * T_j := \sum_{e,f} \Phi_{ije} g^{ef} T_f.$$

It is immediate to check that this *quantum product* $*$ is commutative, with unit T_0. It's far less obvious that $*$ is associative; see the preceding lecture for a proof.

§3. **The projective plane.** As a first example, which is simple yet gives the general flavor, let's work out the above theory for $X = \mathbb{P}^2$. Let T_0, T_1, and T_2 be the classes of X, a line, and a point. Say $\beta = dT_1$ and set $\delta := \frac{1}{2}(d-1)(d-2)$. Then

$$I_\beta(T_2^{n_2}) = \text{the number } N_d \text{ of } \delta\text{-nodal plane curves through } n_2 \text{ points}$$
$$= 0, \text{ unless } n_2 = 3d - 1.$$

For instance, N_1 is simply the number of lines through two points, namely, 1. Remarkably, all the other N_d follow formally from this one! (This fact was noted by Kontsevich first, according to [DF-I, p.82], and it has inspired the determination, via quantum cohomology, of a lot of new geometric numbers.) With the above choice of the T_i, we find

$$\Phi^{cl} = \frac{1}{2}y_0^2 y_2 + \frac{1}{2}y_0 y_1^2, \text{ and } \Gamma = \sum_{d \geq 1} N_d e^{dy_1} \frac{y_2^{3d-1}}{(3d-1)!}.$$

Moreover, we have

$$(g_{ef}) = \begin{pmatrix} 0 & 0 & 1 \\ 0 & 1 & 0 \\ 1 & 0 & 0 \end{pmatrix} = (g^{ef}).$$

Hence, $T_i * T_j = \Phi_{ij2}T_0 + \Phi_{ij1}T_1 + \Phi_{ij0}T_2$. Therefore,

$$T_1 * T_1 = T_2 + \Gamma_{111}T_1 + \Gamma_{112}T_0,$$
$$T_1 * T_2 = \Gamma_{121}T_1 + \Gamma_{122}T_0,$$
$$T_2 * T_2 = \Gamma_{221}T_1 + \Gamma_{222}T_0.$$

Straightforward computation now yields the formulas,

$$(T_1 * T_1) * T_2 = (\Gamma_{222} + \Gamma_{111}\Gamma_{122})T_0 + \ldots$$
$$T_1 * (T_1 * T_2) = \Gamma_{112}\Gamma_{121}T_0 + \ldots$$

So associativity implies the equation,

$$\Gamma_{222} + \Gamma_{111}\Gamma_{122} = \Gamma_{112}\Gamma_{121}.$$

Differentiating Γ, we find

$$\Gamma_{222} = \sum_{d \geq 2} N_d e^{dy_1} \frac{y_2^{3d-4}}{(3d-4)!}, \qquad \Gamma_{122} = \sum_{d \geq 1} d N_d e^{dy_1} \frac{y_2^{3d-3}}{(3d-3)!},$$

$$\Gamma_{112} = \sum_{d \geq 1} d^2 N_d e^{dy_1} \frac{y_2^{3d-2}}{(3d-2)!}, \qquad \Gamma_{111} = \sum_{d \geq 1} d^3 N_d e^{dy_1} \frac{y_2^{3d-1}}{(3d-1)!}.$$

Multiplying, we get

$$\Gamma_{112}^2 = \sum_{d \geq 2} \sum_{d_1+d_2=d} d_1^2 N_{d_1} d_2^2 N_{d_2} e^{dy_1} \frac{y_2^{3d-4}}{(3d_1-2)!(3d_2-2)!},$$

$$\Gamma_{111}\Gamma_{122} = \sum_{d \geq 2} \sum_{d_1+d_2=d} d_1^3 N_{d_1} d_2 N_{d_2} e^{dy_1} \frac{y_2^{3d-4}}{(3d_1-1)!(3d_2-3)!}.$$

Finally, equating coefficients yields Kontsevich's celebrated recursion formula,

$$N_d = \sum_{d_1+d_2=d} N_{d_1} N_{d_2} \left[d_1^2 d_2^2 \frac{(3d-4)!}{(3d_1-2)!(3d_2-2)!} - d_1^3 d_2 \frac{(3d-4)!}{(3d_1-1)!(3d_2-3)!} \right].$$

The following table gives the first five values of the number N_d of rational plane curves of degree d (with δ nodes) through $3d - 1$ points:

d	1	2	3	4	5
N_d	1	1	12	620	87304
δ	0	0	1	3	6
$3d-1$	2	5	8	11	14

The first three values of N_d have been known for a long time. The fourth was found (after a lot of work) by Zeuthen in 1873. The fifth number was found as above by Kontsevich in December 1993 (it had already been found implicitly via more traditional means by Ran and by Vainsencher).

Along the same lines, we can work out the cases of $\mathbb{P}^3$ and of the quadric threefold Q^3 in $\mathbb{P}^4$ (the details are found in [F-P, §9]) and the case of the point-line incidence variety I^3 in $\mathbb{P}^2 \times \mathbb{P}^2$ (the details are found in [DF-I, pp.125–29]). It is also possible to handle conditions of tangency by using the associativity of an appropriate generalized quantum cohomology ring. This was done to some extent by Di Francesco and Itzykson in [DF-I, pp.103–105]) and in full by Ernström and Kennedy in [E-K]. On the other hand, Pandharipande [P] worked with conditions of tangency by using intersection theory on $\overline{M}_{0,n}(\mathbb{P}^r)$ without the power of associativity.

§4. **Feynman diagrams.** Feynman diagrams can be used as an effective mnemonic device for writing down the associativity equations efficiently. The sum,

$$F(i,j|k,l) := \sum_{e,f} \Phi_{ije} g^{ef} \Phi_{fkl},$$

is represented by the diagram,

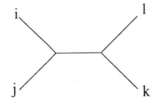

Heuristically, i and j are coupled on the left, and k and l are coupled on the right. Each couple can be in a number of intermediate "states," which are indexed by e and f, and quantified by Φ_{ije} and Φ_{fkl}. Each pair of states is correlated by a "propagator," which is represented by the horizontal link, and quantified by g^{ef}. The total "4-point correlation" is represented by the diagram, and quantified by the above sum $F(i, j|k, l)$.

The same 4-point correlation can be decomposed in a second way into a complete set of intermediate states. The corresponding diagram is

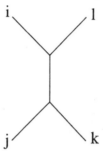

So the corresponding sum is $F(j, k|l, i)$. In physics, the duality relation of topological field theory is symbolized by an equation, with the first diagram above on the left and the second on the right. This duality relation corresponds to the following associativity equation:

$$A(i, j, k, l) : F(i, j|k, l) = F(j, k|l, i).$$

(This differential equation was called a *WDVV equation* after E. Witten, R. Dijkgraaf, H. Verlinde and E. Verlinde by B. Dubrovin.) Every associativity equation arises in this way for suitable values of i, j, k, l. However, many of the equations are equivalent, and others are trivially satisfied, as we'll now see.

The Φ_{ije} are symmetric in i, j, e, and the g^{ef} are symmetric in e, f. So a simple formal calculation yields the following equivalences of equations:

$$A(i, j, k, l) \equiv A(j, k, l, i) \equiv A(k, l, i, j) \equiv A(l, i, j, k) \equiv$$
$$A(i, l, k, j) \equiv A(j, i, l, k) \equiv A(k, j, i, l) \equiv A(l, k, j, i).$$

To obtain each of these eight equations, pick one of the four indices and read progressively around either one of the above diagrams either counterclockwise or clockwise.

Similarly, there are two more groups of eight equivalent equations, and they correspond to the following two duality relations:

If all four of i, j, k, l are distinct, then there are twenty-four possible equations, and they divide into the above three groups of eight equivalent equations. If only three indices are distinct, say $i = j$, then there is, up to equivalence, only one nontrivial associativity equation $A(i, i, k, l)$, and it corresponds to the duality relation,

The equations of the other two groups are trivially satisfied because of the symmetry of the Φ_{ije}. If only two indices are distinct, say $i = j$ and $k = l$, then there is again, up to equivalence, only one nontrivial associativity equation $A(i, i, k, k)$, and it corresponds to the duality relation,

If three or four indices coincide, then the resulting equations are automatically satisfied. The same happens if one of the indices is 0. Indeed, $\Gamma_{0jk} = 0$; so null indices matter only for Φ^{cl}. However, the classical product is already associative!

Consequently, if the rank of $A^*(X)$ is $1 + m$, then the total number of basic associativity equations is

$$3\binom{m}{4} + m\binom{m-1}{2} + \binom{m}{2} = \frac{m(m-1)(m^2 - m + 2)}{8}.$$

For example, for $m = 2, \ldots, 7$, the numbers are 1, 6, 21, 55, 120, 231. Thus, if X is $\mathbb{P}^1 \times \mathbb{P}^1$ or $\mathbb{P}^3$, then m is 3, and the number of basic equations is 6.

Thus, given $A^*(X)$, including the anticanonical class $-K_X$, and a suitable number of initial conditions, it is a formal matter to set up the associativity equations and then to solve for the Gromov–Witten invariants. An algorithm to do so was recently developed and implemented as a (200K) C-program, `farsta`, by Andrew Kresch.

§5. **Surfaces in general.** Again, let X be a smooth irreducible projective variety of any dimension, and suppose that $H_*(X, \mathbb{Z})$ is algebraic. Then in particular $H_{2n+1}(X, \mathbb{Z}) = 0$. So the universal-coefficient theorem implies that $H^{2n+1}(X, \mathbb{Z}) = 0$ and that $H^{2n}(X, \mathbb{Z})$ is torsion-free.

By Hodge theory, $H^1(X, \mathbb{C}) = H^1(\mathcal{O}_X) \oplus H^0(\Omega^1_X)$; so $H^1(\mathcal{O}_X) = 0$. Further,

$$H^2(X, \mathbb{C}) = H^2(\mathcal{O}_X) \oplus H^1(\Omega^1_X) \oplus H^0(\Omega^2_X),$$

and the algebraic cycles map into $H^1(\Omega^1_X)$. They span all of $H^2(X, \mathbb{C})$ by assumption. Hence $H^2(\mathcal{O}_X) = 0$.

Consider the exponential sequence:

$$0 \to \mathbb{Z}_X \to \mathcal{O}_X \to \mathcal{O}^*_X \to 0.$$

It yields the long exact sequence:

$$H^1(\mathcal{O}_X) \to H^1(\mathcal{O}^*_X) \to H^2(X, \mathbb{Z}) \to H^2(\mathcal{O}_X).$$

The extreme terms vanish. So we get

$$\mathrm{Pic}(X) = H^1(\mathcal{O}^*_X) \xrightarrow{\sim} H^2(X, \mathbb{Z}).$$

Thus $\mathrm{Pic}(X)$ is discrete and torsion-free. In particular, if X is a *surface,* and if β is given in $A_1 X := H_2(X, \mathbb{Z})$, then the divisors D of class (Poincaré dual to) β are linearly equivalent.

Let $\mu \colon \mathbb{P}^1 \to X$ be a map that's birational onto its image. Form the sequence,

$$0 \to \mathcal{T}_{\mathbb{P}^1} \to \mu^* \mathcal{T}_X \to \mathcal{N}_\mu,$$

$\big($where $\mathcal{N}_\mu$ is the dual of $\mathrm{Ker}(\mu^* \Omega^1_X \to \Omega^1_{\mathbb{P}^1})\big)$. This sequence must be of the form,

$$0 \to \mathcal{O}(2) \xrightarrow{u} \mathcal{O}(a) \oplus \mathcal{O}(b) \to \mathcal{O}(c),$$

say with $a \geq b$. Since the map u is nonzero, so is its composition with the projection to $\mathcal{O}(a)$ or else with that to $\mathcal{O}(b)$; hence, $a \geq 2$ or $b \geq 2$. Therefore, $a \geq 2$ since $a \geq b$. Similarly, $c \geq a$ or $c \geq b$; hence $c \geq b$. Further, if X is convex, then $b \geq 0$ and so also $c \geq 0$.

Suppose that X is a convex surface. Set $D := \mu_* \mathbb{P}^1$. Then $\mathcal{N}_\mu = \mathcal{O}(\mu^* D)$. So $D^2 \geq 0$ as $c \geq 0$. Furthermore, $\langle -K_X \cdot D \rangle \geq 2$ with $-K_X = c_1 \mathcal{T}_X$ as $a \geq 2$ and $b \geq 0$. In particular, there are no (-1)-curves. Hence X is relatively minimal. Therefore, if X is rational, then $X = \mathbb{P}^2$ or $X = \mathbb{F}_e$ (the Hirzebruch surface). In the latter case, $X = \mathbb{P}^1 \times \mathbb{P}^1$ necessarily, because $\mathbb{F}_e$ has a section of square $-e$. If X is irrational, then K_X is nef and so, since $\langle -K_X \cdot D \rangle \geq 2$, there are no rational curves on X.

In sum, if X is a convex surface with $H_*(X, \mathbb{Z})$ algebraic, then either $X = \mathbb{P}^2$, or $X = \mathbb{P}^1 \times \mathbb{P}^1$ (and in both these cases $X = G/P$), or else X has no rational curves! In other words, for surfaces, the requirement of convexity is rather restrictive for the applications of quantum cohomology to enumerative geometry.

From now on, X is a surface with $H_*(X, \mathbb{Z})$ algebraic, but X is not necessarily convex. Given $\beta \in A_1 X$, set

$$k(\beta) := \langle -K_X \cdot \beta \rangle = \int_\beta c_1(TX).$$

Let π be the class of a point. If $I_\beta(\pi^n) \neq 0$, then the equations,

$$n \cdot \operatorname{codim} \pi = \exp. \dim(\overline{M}) = \dim X + k(\beta) + n - 3,$$

give the formula,

$$n = k(\beta) - 1,$$

because $\operatorname{codim} \pi = 2$ and $\dim X = 2$. Set therefore

$$N_\beta := I_\beta(\pi^{k(\beta)-1}).$$

Normally, N_β is the number of curves of class β with $p_a(\beta)$ nodes (so they're immersed $\mathbb{P}^1$s) that pass through $k(\beta) - 1$ points. Recall that, if β is not effective (that is, not the class of the image of a tree of $\mathbb{P}^1$s), then $N_\beta = 0$; it is reasonable to conjecture that, if the arithmetic genus $p_a(\beta)$ is strictly negative, then again $N_\beta = 0$.

As in §2, replace the potential Φ by the sum $\Phi^{cl} + \Gamma$ where

$$\Gamma(\gamma) := \sum_{\beta \neq 0} N_\beta e^{\int_\beta \gamma} \frac{y_m^{k(\beta)-1}}{(k(\beta)-1)!}.$$

Write $\gamma = \gamma_0 + \gamma_1 + \gamma_2$ with $\gamma_i \in A_i(X)$. Then

$$\Phi^{cl} = \sum_{i+j+k=3} \frac{1}{i!j!k!} \int_X \gamma_0^i \gamma_1^j \gamma_2^k.$$

For the integral to be nonzero, necessarily $j + 2k = 2$, and so either $k = 0$, $j = 2$, $i = 1$ or $k = 1$, $j = 0$, $i = 2$. So

$$\Phi^{cl} = \frac{1}{2} \int \gamma_0 \gamma_1^2 + \frac{1}{2} \int \gamma_0^2 \gamma_2.$$

For $0 < i, j < m$, the duality relation

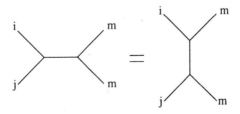

corresponds to the associativity equation,

$$A(i,j,m,m) : \sum_{e,f} \Phi_{ije} g^{ef} \Phi_{fmm} = \sum_{e,f} \Phi_{jme} g^{ef} \Phi_{fmi}.$$

On the left, consider the terms with $f = m$. First, $\Phi^{cl}_{mmm} = 0$; so $\Phi_{mmm} = \Gamma_{mmm}$. Next, $g^{em} = 0$ if $e \neq 0$, and $g^{0m} = 1$. Finally, $\Phi_{ij0} = \Phi^{cl}_{ij0} = g_{ij}$. Hence there's only one nonzero term with $f = m$, and it's equal to $g_{ij}\Gamma_{mmm}$. On the other hand, $\Phi^{cl}_{ije} = 0$ unless $e = 0$, and $\Phi^{cl}_{efm} = 0$ unless $e = 0$ and $f = 0$. Therefore, the associativity equation yields a formula of the following form:

$$g_{ij}\Gamma_{mmm} = \text{a certain quadratic polynomial in the } \Gamma_{efm}.$$

As in the case of $\mathbb{P}^2$, this formula yields a recurrence relation of the following form:

$$g_{ij}N_\beta = \sum_{\beta_1+\beta_2=\beta} N_{\beta_1} N_{\beta_2} * \{ * - * \}.$$

Only one of these recurrence relations is needed to solve for the N_β; however, the others serve to reduce the number of necessary initial conditions, although redundantly. (For a further discussion of this matter whenever A^*X is generated by A^1X, see Kresch's paper [Kresch].)

§6. Del Pezzo and Hirzebruch surfaces.
In this final section, we'll consider two examples: the Del Pezzo surfaces, and the Hirzebruch surfaces.

A *Del Pezzo surface* is a smooth irreducible projective surface X such that $-K_X$ is ample. One such X is $\mathbb{P}^1 \times \mathbb{P}^1$; it is also a Hirzebruch surface, and will be considered below. Otherwise, X is obtained by blowing up the plane at r points in general position, where $0 \leq r \leq 8$. Note that X is not convex for $r > 0$. A natural basis for A^1X is $\{h, e_1, \ldots, e_r\}$ where h is the pullback of the class of a line, and e_i is the class of the ith exceptional divisor. The potential Φ is a well-defined power series because $-K_X$ is ample. So the associativity equation $A(1,1,m,m)$ yields the following recursion formula:

$$N_\beta = \sum N_{\beta_1} N_{\beta_2} \langle \beta_1 \cdot \beta_2 \rangle \langle h \cdot \beta_1 \rangle \left[\langle h \cdot \beta_2 \rangle \binom{k(\beta)-4}{k(\beta_1)-2} - \langle h \cdot \beta_1 \rangle \binom{k(\beta)-4}{k(\beta_2)-1} \right].$$

Needless to say, for $r = 0$, we recover the formula of §3 for the plane. The case $r = 6$, where X is equal to a cubic surface in $\mathbb{P}^3$, was worked out in detail by Di Francesco and Itzykson in [DF-I, §3.3]; the general case was treated briefly by Kontsevich and Manin in [K-M, §5.2.3]. The case of arbitrary r (possibly greater than 8 where $-K_X$ is *not* ample) was treated in depth by Göttsche and Pandharipande in [G-P]; they also considered the enumerative significance of the N_β.

The *Hirzebruch surface* of index e is the rational ruled surface,

$$\mathbb{F}_e := \mathbb{P}\big(\mathcal{O}_{\mathbb{P}^1} \oplus \mathcal{O}_{\mathbb{P}^1}(e)\big).$$

It has a unique section E over $\mathbb{P}^1$ such that $E^2 = -e$ if $e \geq 1$. However, $\mathbb{F}_0$ is $\mathbb{P}^1 \times \mathbb{P}^1$; in this case, let E be any section. Of course, $\mathbb{F}_0$ can be embedded as the quadric surface in $\mathbb{P}^3$, and the latter was studied in some detail by Di Francesco and Itzykson in [DF-I, §3.2]. The next surface $\mathbb{F}_1$ is the blowup of $\mathbb{P}^2$ at a single point, a Del Pezzo again.

For any e, a natural basis for $A^1 X$ is given by the class $[E]$ and that $[F]$ of a fiber. For $e \geq 2$, there are infinitely many classes $\beta = a[E] + b[F]$ having $a, b \geq 0$ and given $k(\beta)$. However, only finitely many have $p_a(\beta) \geq 0$. Indeed,

$$k(\beta) = (2 - e)a + 2b \quad \text{and} \quad p_a(\beta) = (a - 1)(2b - 2 - ae)/2.$$

So, if $p_a(\beta) \geq 0$ and $a \geq 2$, then $2b - ae \geq 2$; hence, if also $k(\beta)$ is given, then a is bounded, and so b is bounded too. So the conjecture of the preceding section would imply that the potential Φ is a well-defined power series, but this question is unimportant when Φ is modified as explained in §2. The conjecture would also provide some useful initial conditions, but these conditions can also be obtained by using a number of associativity equations.

The associativity equation $A(1, 2, 3, 3)$ yields the recursion formula,

$$N_\beta = \sum N_{\beta_1} N_{\beta_2} \langle \beta_1 \cdot \beta_2 \rangle \langle E \cdot \beta_1 \rangle \left[\langle F \cdot \beta_2 \rangle \binom{k(\beta) - 4}{k(\beta_1) - 2} - \langle F \cdot \beta_1 \rangle \binom{k(\beta) - 4}{k(\beta_2) - 1} \right].$$

(This formula was worked out for the first time by Ragni Piene and the lecturer in March of 1994.) It turns out experimentally that, on writing $N(a, b; e)$ for N_β where $\beta = a[E] + b[F]$ on $\mathbb{F}_e$, we find the relation,

$$N(a, b; e) = N(a, b + a; e + 2).$$

A conceptual explanation for it (explained to the lecturer by Sheldon Katz in September 1994) is this: $\mathbb{F}_e$ degenerates into $\mathbb{F}_{e+2}$, transforming F to F and E to $E + F$, while leaving the quantum cohomology invariant.

The enumerative significance of the N_β is, of course, nontrivial to establish. On the other hand, there are more geometric computations of related numbers, which do not make use of quantum cohomology: Caporaso and Harris have obtained numbers of *irreducible* rational curves of class β on $\mathbb{F}_1$, $\mathbb{F}_2$, $\mathbb{F}_3$ and of a special β on an arbitrary $\mathbb{F}_e$. Following in their footsteps, Vakil has obtained all the numbers of both the reducible and irreducible curves on an arbitrary $\mathbb{F}_e$ (and the corresponding numbers in arbitrary genus as well!). Abramovich and Bertram, in work in progress, have been obtaining numbers by carefully studying the degeneration of $\mathbb{F}_e$ into $\mathbb{F}_{e+2}$.

Here is a concrete example, which illustrates some of the subtleties involved in the enumeration. (This example was explained to the lecturer by Dan

Abramovich in November 1995.) Consider $\mathbb{F}_2$. Note that $k(a[E] + b[F]) = 2b$, which is independent of a! Now, take $\beta := 3[E] + 6[F]$. Then $k(\beta) - 1 = 11$ and

$$N_\beta = 3510 = 2232 + 2 \cdot 636 + 6 \cdot 1 + 0$$

where the four terms on the right arise as follows:

2232   is the contribution of the irreducible curves of class β through 11 general points.

$2 \cdot 636$ is the contribution of the curves breaking up as E union an irreducible curve of class γ, where $\gamma := 2[E] + 6[F]$, through the 11 points. The latter curve meets E twice: $\langle (2E + 6F) \cdot E \rangle = -4 + 6 = 2$. So there are two ways to partially normalize the curve into a tree of $\mathbb{P}^1$s. There are 636 such curves of class γ, because $N_\gamma = 640$ and $N_\gamma = 636 + 4 \cdot 1$, where $4 \cdot 1$ is the contribution (to N_γ now!) of curves breaking up as E union a curve of class $\delta := [E] + 6[F]$. Indeed, $\langle \delta \cdot [E] \rangle = -2 + 6 = 4$, and therefore there are 4 ways to attach a curve of class δ to E. Finally, there are N_δ curves of class δ through the 11 points, and $N_\delta = 1$ because $p_a(\delta) = 0$.

$6 \cdot 1$ is the contribution of curves splitting as $2[E] + \delta$ where $\delta := [E] + 6[F]$ as before. The two reduced components again meet in 4 points as the intersection number $\langle (E + 6F) \cdot E \rangle$ is 4. There are 6 ways to map a tree of 3 $\mathbb{P}^1$s with 2 nodes onto each curve, mapping the two extreme $\mathbb{P}^1$s to E, the connecting $\mathbb{P}^1$ to the other component, and the 2 nodes to 2 of the 4 points. Finally, as before, there is 1 curve of class δ through the 11 points.

0   is the contribution of the same curves as in the previous case, but with a tree of 2 $\mathbb{P}^1$s with 1 node and 1 $\mathbb{P}^1$ mapping 2-to-1 to E. It can be shown that these form a component of $\overline{M}$ of the wrong dimension and contribution 0, via an explicit analysis of the degeneration of $\mathbb{F}_e$ into $\mathbb{F}_{e+2}$.

---

6. About $\overline{M}(X, \beta)$—I. Ciocan-Fontanine
October 22, 1996

We go back now to the proof of the properties of $\overline{M}_{0,n}(X, \beta)$ that we used in the enumerative geometry computations and the construction of QH^*. The plan for this lecture is:

—§1. *Overview of the construction of $\overline{M}_{g,n}(X, \beta)$ and of the proofs*
—§2. *The idea behind the construction for $X = \mathbb{P}^r$*
—§3. *Begin the formal proofs*

§1. **Overview.** Let X be a projective variety over $\mathbb{C}$, $\beta \in A_1 X$. We have a functor

$$\overline{\mathcal{M}}_{g,n}(X, \beta): \quad \{\text{schemes}/\mathbb{C}\} \to \{\text{Sets}\}$$

defined by

$$\mathcal{M}_{g,n}(X, \beta)(S) = \{\text{isom. classes of stable maps of genus } g, \, n\text{-pointed curves},$$
$$\text{etc.}\}$$

We first list the results we need.

THEOREM 1. *There exists a projective coarse moduli space* $\overline{M}_{g,n}(X, \beta)$ *for this functor.*

The technical formulation of this statement is: there exists a projective scheme $\overline{M}_{g,n}(X, \beta)$ together with a natural transformation of functors

$$\phi : \overline{\mathcal{M}}_{g,n}(X, \beta) \longrightarrow \text{Hom}(\cdot, \overline{M}_{g,n}(X, \beta))$$

satisfying

(1) $\phi(\text{Spec}(\mathbb{C}))$ is a bijection of sets;
(2) If there is a $\psi : \overline{\mathcal{M}}_{g,n}(X, \beta) \longrightarrow \text{Hom}(\cdot, Z)$, then there is a unique morphism $\overline{M}_{g,n}(X, \beta) \xrightarrow{\gamma} Z$ such that $\psi = \tilde{\gamma} \circ \phi$, with $\tilde{\gamma} = \text{Hom}(\cdot, \gamma)$.

If we assume in addition that X is nonsingular and convex, and that $g = 0$, then we can say more: we have a local description of $\overline{M}_{0,n}(X, \beta)$:

THEOREM 2.

(i) $\overline{M}_{0,n}(X, \beta)$ *is a locally normal projective variety of pure dimension*

$$\dim X + \int_\beta c_1(T_X) + n - 3$$

(ii) $\overline{M}_{0,n}(X, \beta)$ *is locally a quotient of a smooth quasi-projective variety by a finite group;*
(iii) $\overline{M}^*_{0,n}(X, \beta)$ *(=automorphism-free locus) is smooth, with a universal family.*

The *boundary* of $\overline{M}_{0,n}(X, \beta)$ is the complement of the subset parametrizing irreducible curves; that is, $\overline{M}_{0,n}(X, \beta) - M_{0,n}(X, \beta)$.

THEOREM 3. *The boundary of* $\overline{M}_{0,n}(X, \beta)$ *is a normal crossing divisor, up to a quotient by a finite group.*

Outline of the proofs:

(1) *Existence:* First construct $\overline{M}_{g,n}(\mathbb{P}^r, d)$ and show that this is a projective scheme. Next, for X projective choose an embedding $X \xrightarrow{i} \mathbb{P}^r$, and show that there exists a natural closed subscheme $\overline{M}_{g,n}(X, d) \subset \overline{M}_{g,n}(\mathbb{P}^r, d)$; then $\overline{M}_{g,n}(X, d) = \amalg_{i_*\beta=d(\text{line})} \overline{M}_{g,n}(X, \beta)$. The universality property of the coarse moduli space will imply the independence of the space from the chosen embedding.
(2) *Local structure:* Comes for free from the construction when $X = \mathbb{P}^r$.
(3) *Boundary:* Ditto.

§2. Outline for $X = \mathbb{P}^r$. Let $d, r > 0$, and let $(C; \{p_i\}, \mu)$ be a stable n-pointed map to $\mathbb{P}^r$, with image of degree d.

Choose coordinates $(x_0 : \cdots : x_r)$ for $\mathbb{P}^r$. The map $(C; \{p_i\}, \mu)$ determines and is determined by the data of

$(C, \{p_i\})$, n-pointed quasi-stable;
a line bundle $\mathcal{L}$ on C, that is $\mathcal{L} = \mu^* \mathcal{O}_{\mathbb{P}^r}(1)$;
$r + 1$ general sections $s_i = \mu^*(x_i)$.

This is what we would like to parametrize.

A generic map (if there is one) will have transversal intersection with the coordinate hyperplanes $\{x_i = 0\}$, away from $\{p_i\}$ and from the nodes. The divisor of s_i consists then of distinct points $\{q_{i1}, \ldots, q_{id}\}$: we get additional $d(r+1)$ marked points on C. Assume C is generic:

CLAIM. $((r, d) \neq (1, 1))$ μ is stable if and only if $(C; \{p_i\}, \{q_{ij}\})$ is Deligne-Mumford-Knudsen stable.

PROOF. $\Longrightarrow$: Say that $E \subset C$, $E \cong \mathbb{P}^1$; if $\mu(E)$ is a point, we are done by the stability of μ. Otherwise, there are two cases: if $E = C$, all $d(r+1) \geq 3$ extra marked points are on C; and if $E \neq C$, we have at least one node, and $(r+1) \geq 2$ of the new markings on E. So every contracted component has at least 3 special points, as needed.

The other direction is easier. $\square$

Summing up, the generic $(C, \{p_i\}, \mu)$ determines a point in $\overline{M}_{g,m}$, with $m = n + d(r+1)$. Note that there is a $\mathbb{C}^*$ ambiguity in the choices of each of the s_i, modulo a $\mathbb{C}^*$ by homogeneity. That is, given the points corresponding to the sections (the information in the q_{ij}'s) there is still a $(\mathbb{C}^*)^r$ worth of additional choices to be made.

Further, we have to take account of the permutations of the points in each hyperplane section, that is, of the action of $G = S_d \times \cdots \times S_d = (S_d)^{r+1}$, where the i-th factor permutes $\{q_{i1}, \ldots, q_{id}\}$.

What about nongeneric curves? All curves are generic for some choice of the coordinates; a given choice gives a coordinate patch for an open in a cover of $\overline{M}$. Then we need to show that these coordinate patches do patch, and a boundedness result to show that the scheme is of finite type. Then we need to show that the scheme is proper; and finally that it is projective (this is technically harder).

Summarizing: we have to consider a (quotient of a) torus bundle over a subset of $\overline{M}_{g,m}$. Which subset? (In the genus-0 case, we will end up with an open subset of $\overline{M}_{0,m}$.) Let $B \subset \overline{M}_{g,m}$ be the subset determined as above. Remark: on C, the r divisors divs_i are linearly equivalent; this puts a condition on $\overline{M}_{g,m}$, as we need the divisors $\{q_{01}, \ldots, q_{0d}\}$, $\ldots$, $\{q_{r1}, \ldots, q_{rd}\}$ on $C \in B$ to be linearly equivalent divisors (and very ample). In particular, the degrees of these divisors on a given component must be equal; in genus 0, this suffices essentially to determine everything, as the number of points in a divisor determines its class. In genus> 0, the ambiguity is measured by the Jacobian; in practice, we can't get our hands on the local structure of the resulting spaces.

§3. **Formal proofs.** Say $\mathbb{P}^r = \mathbb{P}V$, where V is an $(r+1)$-dimensional complex vector space; $H^0(\mathcal{O}(1)) = V^*$ Fix $\bar{t} = \{t_0, \ldots, t_r\}$, a basis of V^*.

DEFINITION 1. A $\bar{t}$-stable family of degree-d maps from n-pointed curves to $\mathbb{P}^r$ consists of

$$(\pi : \mathcal{C} \to S; \{p_i\}_{i=1,\ldots,n}, \{q_{ij}\}_{0 \le i \le r, 1 \le j \le d}, \mu)$$

such that

(i) $(\pi : \mathcal{C} \to S, \{p_i\}, \mu)$ is a stable family of degree-d maps from n-pointed curves to $\mathbb{P}^r$;

(ii) $(\pi : \mathcal{C} \to S; \{p_i\}, \{q_{ij}\})$ is a stable $m = n + d(r+1)$-pointed curve;

(iii) $\mu^*(t_i) = q_{i1} + \cdots + q_{id}$ as effective Cartier divisors.

DEFINITION 2. $\overline{\mathcal{M}}_{g,n}(\mathbb{P}^r, d, \bar{t})$ is the functor of $\bar{t}$-rigid stable maps: that is, $\overline{\mathcal{M}}_{g,n}(\mathbb{P}^r, d, \bar{t})(S) = \{$isomorphism classes of families over S as in Definition 1$\}$.

PROPOSITION 1. *There exists a quasi-projective moduli space* $\overline{M}_{g,n}(\mathbb{P}^r, d, \bar{t})$, *which is coarse for* $g > 0$ *and fine and nonsingular for* $g = 0$.

PROOF. (Only for $g = 0$.) Let $\overline{M} = \overline{M}_{0,m}$; $\pi : \overline{U} \to \overline{M}$ the universal family; p_i, q_{ij} are sections of $\overline{U} \to \overline{M}$. As $\overline{U}$ is nonsingular, $q_{i1} + \cdots + q_{id}$ determines a line bundle $\mathcal{H}_i = \mathcal{O}_{\overline{U}}(q_{i1} + \cdots + q_{id})$. Denote by s_i a corresponding section, that is a 'global equation' for $q_{i1} + \cdots + q_{id}$.

Denote by $B \subset \overline{M}$ the open subscheme determined by the property that for $b \in B$, the geometric fiber $C_b = \overline{U}_b$ will have $\deg \mathcal{H}_i = \deg \mathcal{H}_j$ on every component of C_b.

$$\begin{array}{ccc} \overline{U}_B & \xrightarrow{\bar{i}} & \overline{U} \\ \pi_B \downarrow & & \downarrow \pi \\ B & \xrightarrow{i} & \overline{M} \end{array}$$

(Note: for $g > 0$, we would require instead that $\mathcal{H}_i \otimes \mathcal{H}_0^{-1}$ 'comes from the base' for all i.) Now $(\pi_B)_* \bar{i}^* (\mathcal{H}_i \otimes \mathcal{H}_0^{-1})$ is a line bundle $\mathcal{G}_i$ on B, and has an associated $\mathbb{C}^*$-bundle $\tau_i : Y_i \to B$ obtained by deleting the zero-section.

Let then $Y = Y_1 \times_B \cdots \times_B Y_r$, coming with projections $\rho_i : Y \to Y_i$, $\tau : Y \to B$. Note that $\tau_i^* \mathcal{G}_i$ is canonically trivial.

$$\begin{array}{ccccc} \overline{U}_Y & \xrightarrow{\bar{\tau}} & \overline{U}_B & \xrightarrow{\bar{i}} & \overline{U} \\ \pi_Y \downarrow & & \pi_B \downarrow & & \downarrow \pi \\ Y & \xrightarrow{\tau} & B & \xrightarrow{i} & \overline{M} \end{array}$$

CLAIM. *On $\overline{U}_Y$ there is a canonical isomorphism $\overline{\tau}^*i^*\mathcal{H}_i \cong \overline{\tau}^*i^*\mathcal{H}_0 =: \mathcal{L}$.*

PF. $\overline{\tau}^*\overline{i}^*\mathcal{H}_i \otimes \mathcal{H}_0^{-1} = \overline{\tau}^*\pi_B^*\pi_{B*}\overline{i}^*\mathcal{H}_i \otimes \mathcal{H}_0^{-1}$ (since $\overline{i}^*\mathcal{H}_i \otimes \mathcal{H}_0^{-1}$ comes from B, $\pi_B^*\pi_{B*}$ leaves it alone), which equals $\pi_Y^*\tau^*\mathcal{G}_i = \pi_Y^*\rho_i^*\tau_i^*\mathcal{G}_i$, and $\tau_i^*\mathcal{G}_i$ is canonically trivial. □

Now $\overline{\tau}^*s_i$ generates $\overline{\tau}^*\mathcal{H}_i$, for $i = 0, \dots, r$. There exists a unique $\mu : \overline{U}_Y \to \mathbb{P}^r$ such that $\mu^*\mathcal{O}(1) = \mathcal{L}$ and $\mu^*t_i = \overline{\tau}^*s_i$.

CLAIM. $(\pi_Y : \overline{U}_Y \to Y, \{p_i\}, \{q_{ij}\}, \mu)$ *is a universal family over Y.*

That is, Y represents the functor of $\overline{t}$-rigid stable maps.

PF. By construction it is a $\overline{t}$-rigid stable family. To show it is universal: pick another $(\pi_S : \mathcal{C}_S \to S, \{p_i\}, \{q_{ij}\}, \nu)$, and show that there is a unique morphism $S \to Y$ such that this family is canonically isomorphic to the pull-back of $\overline{U}_Y$. For this, $\pi_S : \mathcal{C}_S \to S$ is a family of m-pointed curves, in particular, so there is a unique $\lambda : S \to \overline{M}$:

$$
\begin{array}{ccc}
\mathcal{C}_S & \xrightarrow{\overline{\lambda}} & \overline{U} \\
\pi_S \downarrow & & \downarrow \pi \\
S & \xrightarrow{\lambda} & \overline{M}
\end{array}
$$

$\lambda^*\mathcal{H}_i \cong \nu^*\mathcal{O}_{\mathbb{P}^r}(1)$, so $S \xrightarrow{\lambda} \overline{M}$ factors $S \xrightarrow{\lambda} B \to \overline{M}$. There is a canonical isomorphism:

$$
\mathcal{O}_S \cong \pi_{S*}\overline{\lambda}^*\mathcal{H}_i \otimes \mathcal{H}_0^{-1} \cong \lambda^*\mathcal{G}_i \quad ;
$$

because of this, there is a canonical isomorphism $\pi_{S*}\pi_S^*\mathcal{N} \cong \mathcal{N}$ for all line bundles $\mathcal{N}$ on S; in particular, there are canonical isomorphisms

$$
\lambda^*\mathcal{G}_i \cong \pi_{S*}\pi_S^*\lambda^*\mathcal{G}_i = \pi_{S*}\overline{\lambda}^*\pi_S^*\mathcal{G}_i = \pi_{S*}\overline{\lambda}^*\pi_B^*\pi_{B*}\overline{i}^*(\mathcal{H}_i \otimes \mathcal{H}_0^{-1})
$$
$$
\cong \pi_{S*}\overline{\lambda}^*(\mathcal{H}_i \otimes \mathcal{H}_0^{-1})
$$

Therefore there is a canonical $S \to Y$ commuting in

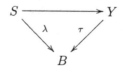

The pull-back of the universal family is the given family on S because at each step all choices were canonical. □

7. The construction of $\overline{M}_{0,n}(\mathbb{P}^r, d)$—J. Thomsen
November 5, 1996

We fix n, r, d, and only deal with the genus= 0 case. Let $V^* = H^0(\mathbb{P}^r, \mathcal{O}_{\mathbb{P}^r}(1))$, and let $\bar{t} = \{t_0, \ldots, t_r\}$ be a basis of V^*. Consider the functor

$$\mathcal{M}_{0,n}(\mathbb{P}^r, d, \bar{t}) : \quad \text{Alg. schemes over } \mathbb{C} \to \text{Sets}$$

sending S to the set of isomorphism classes of $\bar{t}$-rigid families over S. Here a $\bar{t}$-*rigid family* over S is an object

$$(\pi : C \to S; \{p_i\}_{i=1,\ldots,n}, \{q_{ij}\}_{0 \leq i \leq r, 1 \leq j \leq d}, \mu : C \to \mathbb{P}^r)$$

where

(i) $(\pi : C \to S, \{p_i\}, \mu)$ is a stable family of degree-d maps from n-pointed genus-0 curves to $\mathbb{P}^r$;

(ii) $(\pi : C \to S, \{p_i\}, \{q_{ij}\})$ is a stable family of $m = (n + d(r + 1))$-pointed genus-0 Deligne-Mumford stable curves;

(iii) $\mu^*(t_i) = q_{i1} + \cdots + q_{id}$, $i = 0, \ldots, r$.

Recall from Ciocan-Fontanine's lecture:

THEOREM. *There exists a quasiprojective nonsingular variety $\overline{M}(\bar{t})$, which is a fine moduli space with respect to $\overline{\mathcal{M}}_{0,n}(\mathbb{P}^r, d, \bar{t})$.*

Our task is to patch these moduli spaces together into a space $\overline{M} = \overline{M}_{0,n}(\mathbb{P}^r, d)$.

Suppose that we already have a fine moduli space $\overline{M}$ for $\overline{\mathcal{M}}_{0,n}(\mathbb{P}^r, d)$. For a $\bar{t}$-rigid family $(\pi : C \to S, \{p_i\}, \{q_{ij}\}, \mu)$, we consider the diagram

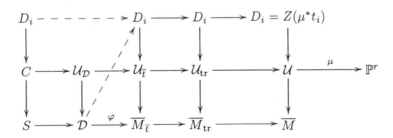

The map $S \to \overline{M}$ realizing the family via pull-back is not surjective, as the curves over S intersect the D_i transversally (by rigidity): so the map must factor through the open subset

$$\overline{M}_{\mathrm{tr}} = \{\text{curves intersecting the } D_i\text{'s transversally}\}$$

Similarly, the map factors through the smaller open $\overline{M}_{\bar{t}}$ where the p_i's are distinct from the D_i's, and the D_i's intersect trivially.

Next, the q_{ij} define a map from S to

$$\mathcal{D} = \prod_i \left[\left(\underbrace{D_i \times_{\overline{M}_{\bar{t}}} \cdots \times_{\overline{M}_{\bar{t}}} D_i}_{d} \right) - \text{big diagonals} \right]$$

through which the map to $\overline{M}_{\bar{t}}$ also factors. As $\mathcal{D}$ is independent of the $\bar{t}$-rigid family we started with, it is a concrete realization of $\overline{M}(\bar{t})$.

Also, $G_{d,r} = \underbrace{S_d \times \cdots \times S_d}_{r+1}$ acts on $\mathcal{D}$ over $\overline{M}_{\bar{t}}$; and we claim that

$$\mathcal{D}/G_{d,r} \xrightarrow{\phi/G_{d,r}} \overline{M}_{\bar{t}}$$

is an isomorphism. To see this (at least on closed fibers), look at the fiber over $s \in S \to \overline{M}_{\bar{t}}$:

$$
\begin{array}{ccc}
\pi_i^{-1}(s) & \longrightarrow & D_i \\
\downarrow & & \downarrow{\scriptstyle \pi_i} \\
\{s\} & \longrightarrow & \overline{M}_{\bar{t}}
\end{array}
\qquad
\begin{array}{ccc}
\varphi^{-1}(s) & \longrightarrow & \mathcal{D} \\
\downarrow & & \downarrow{\scriptstyle \varphi} \\
\{s\} & \longrightarrow & \overline{M}_{\bar{t}}
\end{array}
$$

Clearly $\pi_i^{-1}(s) = d$ points of intersection of t_i with (the curve represented by) s; and

$$\varphi^{-1}(s) = \left\{ (x_{01}, \ldots, x_{0d}, \ldots, x_{r1}, \ldots, x_{rd}) \mid x_{ij} \text{ distinct, and } \{x_{ij}\} = \pi_i^{-1}(s) \right\}$$

Summarizing, the open $\overline{M}_{\bar{t}}$ of $\overline{M}$ (if the latter exists) must be isomorphic to $\mathcal{D}/G_{d,r} = \overline{M}(\bar{t})/G_{d,r}$. So we should be able to construct $\overline{M}_{0,n}(\mathbb{P}^r, d)$ by gluing together $\overline{M}(\bar{t})/G_{d,r}$, where $\bar{t}$ runs through all bases of V^*.

Now

$$\overline{M}(\bar{t})/G_{d,r} \leftrightarrow \begin{array}{l} \text{curves (i.e., points on } \overline{M}_{0,n}(\mathbb{P}^r, d)) \text{ which the hyperplanes } t_i \\ \text{intersect transversally in distinct points and away from the} \\ \text{marked points} \end{array}$$

If $\bar{t}'$ is another basis,

$$\text{``}\overline{M}(\bar{t})/G_{d,r} \cap \overline{M}(\bar{t}')/G_{d,r}\text{''} \leftrightarrow \begin{array}{l} \text{curves which the hyperplanes } t_i \text{ and } t_i' \\ \text{intersect transversally in distinct points (each)} \\ \text{and away from marked points} \end{array}$$

In order to control the patching, we have to construct this intersection explicitly (as we did for $\overline{M}(\bar{t})/G_{d,r}$) a moment ago). This will be obtained as the quotient by $G_{d,r} \times G_{d,r}$ of a suitable $\overline{M}(\bar{t}, \bar{t}')$. In fact it is useful and not harder to consider arbitrary (finite) sets of bases at once:

DEFINITION. Let $\{\bar{t}_\ell\}_{1\le\ell\le h}$ be a set of bases of V^*. Then

$$(\pi : C \to S, \{p_i\}_{1\le i\le n}, \{q_{ij\ell}\}_{0\le i\le r,1\le j\le d,1\le\ell\le h}, \mu)$$

is called a $\{\bar{t}_\ell\}_{1\le\ell\le h}$-*rigid family* if for all ℓ,

$$(\pi : C \to S, \{p_i\}_{1\le i\le n}, \{q_{ij\ell}\}_{0\le i\le r,1\le j\le d}, \mu)$$

is $\bar{t}_\ell$-rigid.

DEFINITION. Define a functor $\overline{\mathcal{M}}(\bar{t}_1, \ldots, \bar{t}_h)$ from Algebraic Schemes to Sets sending S to the set of isomorphism of $\{\bar{t}_\ell\}_{1\le\ell\le h}$-rigid families over S (where isomorphisms of families, etc. are defined in the usual way).

THEOREM. *There is a fine moduli space $\overline{M}(\bar{t}_1, \ldots, \bar{t}_h)$ for this functor, carrying a $G_{d,r} \times \cdots \times G_{d,r}$-action.*

The idea of course is that the intersection "$\cap_{\ell=1}^h \overline{M}(\bar{t}_\ell)/G_{d,r}$" should correspond to $\overline{M}(\bar{t}_1, \ldots, \bar{t}_h)/G_{d,r} \times \cdots \times G_{d,r}$.

PROOF OF THE THEOREM. This is done by induction on h; the case $h = 1$ has already been done. For $h > 1$, assume $\overline{M}(\bar{t}_1, \ldots, \bar{t}_{h-1})$ has been constructed already, let $\bar{t}_h = (t_{h0}, \ldots, t_{hr})$ and consider

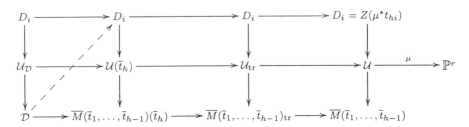

(i) $\overline{M}(\bar{t}_1, \ldots, \bar{t}_{h-1})_{\mathrm{tr}} \subset \overline{M}(\bar{t}_1, \ldots, \bar{t}_{h-1})$ is the maximal open subscheme where the sheaf of relative differentials $\Omega_{D_i/\overline{M}(\bar{t}_1,\ldots,\bar{t}_{h-1})} = 0$;

(ii) $\overline{M}(\bar{t}_1, \ldots, \bar{t}_{h-1})(\bar{t}_h) \subset \overline{M}(\bar{t}_1, \ldots, \bar{t}_{h-1})_{\mathrm{tr}}$ is the maximal open subscheme where the p_i and the D_i are distinct;

(iii) $\mathcal{D} = \prod_i((D_i \times_M \cdots \times_M D_i) - \text{big diag.})$ where $M = \overline{M}(\bar{t}_1, \ldots, \bar{t}_{h-1})(\bar{t}_h)$.

The natural projections $\mathcal{D} \to D_i$ give the new sections q_{ijh}; $\overline{M}(\bar{t}_1, \ldots, \bar{t}_h) := \mathcal{D}$ is the fine moduli space for $\overline{\mathcal{M}}(\bar{t}_1, \ldots, \bar{t}_h)$. The action of the group is the natural one. $\square$

Note that there will be natural maps

$$\overline{M}(\bar{t}_1, \ldots, \bar{t}_h) \xrightarrow{\varphi_{\bar{t}_1,\ldots,\bar{t}_{h-1}\bar{t}_h}} \overline{M}(\bar{t}_1, \ldots, \bar{t}_{h-1})(\bar{t}_h) \overset{\text{open}}{\hookrightarrow} \overline{M}(\bar{t}_1, \ldots, \bar{t}_{h-1})$$

and

(1) $\varphi_{\bar{t}_1,\dots,\bar{t}_{h-1},\bar{t}_h}$ is $(G_{d,r})_1 \times \cdot \times (G_{d,r})_{h-1}$-equivariant;

(2) $\overline{M}(\bar{t}_1,\dots,\bar{t}_h)/(G_{d,r})_h \cong \overline{M}(\bar{t}_1,\dots,\bar{t}_{h-1})(\bar{t}_h)$

Now we can glue the spaces together. The data of the gluing:

• $\overline{M}_{\bar{t}} = \overline{M}(\bar{t})/G_{d,r}$, $\bar{t}$ running through all bases of V^*;

• $\overline{M}_{\bar{t}\bar{t}'}$, defined as $\overline{M}(\bar{t})(\bar{t}')/G_{d,r}$;

• $\varphi_{\bar{t}\bar{t}'} : \overline{M}_{\bar{t}\bar{t}'} \to \overline{M}_{\bar{t}'\bar{t}}$, isomorphisms induced from

$$\begin{array}{ccc}
\overline{M}(\bar{t},\bar{t}')/G_{d,r} \times G_{d,r} & \xrightarrow{\;\sim\;} & \overline{M}(\bar{t}',\bar{t})/G_{d,r} \times G_{d,r} \\
\downarrow & & \downarrow \\
\overline{M}(\bar{t})(\bar{t}')/G_{d,r} = \overline{M}_{\bar{t}\bar{t}'} & \xrightarrow{\;\varphi_{\bar{t}\bar{t}'}\;} & \overline{M}_{\bar{t}'\bar{t}} = \overline{M}(\bar{t}')(\bar{t})/G_{d,r}
\end{array}$$

Compatibility is checked on triple intersection. The resulting scheme is $\overline{M}_{0,n}(\mathbb{P}^r, d)$.

Next, we will check that $\overline{M}_{0,n}(\mathbb{P}^r, d)$ is of finite type over $\mathbb{C}$.

Let $S \in \overline{M}_{0,n}(\mathbb{P}^r, d)$, and let $(C; \{p_i\}, \mu)$ be a corresponding curve. Define

$$\mathcal{L} = \omega_C(p_1 + \dots + p_n) \otimes \mu^*(\mathcal{O}_{\mathbb{P}^r}(3))$$

CLAIM. $\mathcal{L}$ is ample.

Indeed, let $E \subset C$ be a component. Then

(i) $(\omega_C)_{|E} = \omega_E \otimes \mathcal{O}(\text{nodes of } C \text{ along } E)$;

(ii) $\deg(\omega_C(p_1 + \dots + p_n)_{|E}) = -2 + \#$ special points on E;

(iii) $\deg(\mathcal{L}_{|E}) = -2 + \#$ special points on $E + 3\,d_E > 0$ (where $d_E =$ degree of $\mu_*[E]$)

as needed. One can in fact check that

CLAIM. $\mathcal{L}^2$ is very ample, and $h^1(C, \mathcal{L}^2) = 0$.

Define

$$e := \deg \mathcal{L}^2 = 2 \deg \mathcal{L} = 2(-2 + n + 3d)$$

then it follows from Riemann-Roch that $h^0(C, \mathcal{L}^2) = e + 1$. Note that e is independent of C.

Denote by $i : C \hookrightarrow \mathbb{P}^e$ the embedding induced by $\mathcal{L}^2$, and let $\gamma = (i, \mu) : C \hookrightarrow \mathbb{P}^e \times \mathbb{P}^r$. The image of C in $\mathbb{P}^e \times \mathbb{P}^r$ has bidegree (e, d).

Look then at the Hilbert scheme H of genus-0 curves in $\mathbb{P}^e \times \mathbb{P}^r$, of bidegree (e, d); this comes with a universal family $W \to H$. Denote by H_n the space $\underbrace{W \times_H \cdots \times_H W}_{n}$; H_n is a fine moduli space for families of curves of genus 0 and bidegree (e, d), together with n sections. It has a universal family $\mathcal{U}_n$:

$$
\begin{array}{ccccc}
C & \longrightarrow & \mathcal{U}_n & \longrightarrow & \mathbb{P}^e \times \mathbb{P}^r \times H_n & \longrightarrow & \mathbb{P}^r \\
\downarrow & & \downarrow & & & & \\
\mathrm{Spec}\mathbb{C} & \longrightarrow & H_n & & & &
\end{array}
$$

The space H_n is of finite type over $\mathcal{C}$; we want to deduce that $\overline{M}_{0,n}(\mathbb{P}^r, d)$ is then also of finite type.

Given a basis $\bar{t}$ of V^*, we let $H_{n,\bar{t}} \subset H_n$ be the open subscheme where the elements of the basis intersect the curve transversally and in distinct points. Let $H_b = \cup_{\bar{t} \text{ basis of } V^*} H_{n,\bar{t}}$.

(i) H_b is of finite type, so H_b can be covered by a finite number of $H_{n,\bar{t}}$;
(ii) by Bertini's theorem, every stable map from n-pointed genus-0 curves to $\mathbb{P}^r$ is induced from H_b.

The conclusion is that there exist a finite number $\bar{t}_1, \ldots, \bar{t}_h$ of bases of V^* such that

$$
\overline{M}_{0,n}(\mathbb{P}^r, d) = \cup_{i=1}^{h} \overline{M}(\bar{t}_i)
$$

The space $\overline{M}_{0,n}(\mathbb{P}^r, d)$ is then of of finite type, since it is covered by finitely many schemes of finite type.

---

8. $\overline{M}_{0,n}(X, \beta)$—E. Tjøtta
November 12, 1996

Our goal:

THEOREM 1. *If X is projective, there exists a coarse moduli space $\overline{M}_{0,n}(X, \beta)$.*

THEOREM 2. *For X nonsingular and convex:*

i) $\dim \overline{M}_{0,n}(X, \beta) = \dim X + \int_\beta c_1(TX) + n - 3$;
ii) $\overline{M}_{0,n}(X, \beta)$ *is locally a quotient of a nonsingular variety by a finite group.*

Recall how this was done for $X = \mathbb{P}^r$: we constructed a fine moduli space $\overline{M}_{0,n}(\mathbb{P}^r, d, \bar{t})$ for $\bar{t}$-rigid maps, where $\bar{t} = \{t_0, \ldots, t_r\}$ is a basis for $H^0(\mathbb{P}^r, \mathcal{O}(1))$. This space $\overline{M}_{0,n}(\mathbb{P}^r, d, \bar{t})$ is a $(\mathbb{C}^*)^r$-bundle over an open B in $\overline{M}_{0,m}$, with $m = n + d(r+1)$. The group $G = (S_d)^{r+1}$ acts on $\overline{M}_{0,n}(\mathbb{P}^r, d, \bar{t})$, and the quotients as $\bar{t}$ varies glue together, giving $\overline{M}_{0,n}(\mathbb{P}^r, d)$.

A) $\overline{M}_{0,n}(X, \beta)$. Let X be a subvariety of $\mathbb{P}^r$: $X \xrightarrow{i} \mathbb{P}^r$; let $i_*\beta = d[\text{line}]$. There exists a closed subscheme $\overline{M}_{0,n}(X, \beta, \bar{t}) \subset \overline{M}_{0,n}(\mathbb{P}^r, d, \bar{t})$ such that for all families of $\bar{t}$-rigid stable maps $\mu : \mathcal{C} \to \mathbb{P}^r$ over a base S, $\exists \tilde{\mu} \iff \exists \tilde{\nu}$ in the commutative diagram:

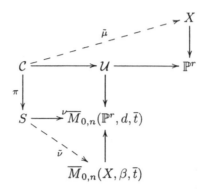

(with $\mathcal{U}$ the universal family, and sections $S \to \mathcal{C}$, etc., as usual).

PROOF. For $\mu : \mathcal{C} \to X$ and $k > 0$, $H^1(\mathcal{C}, \mu^*\mathcal{O}(k)) = 0$; by base change, $\pi_*\mu^*\mathcal{O}(k)$ is a vector bundle with fibers $\pi_*\mu^*\mathcal{O}(k)_s = H^0(\mathcal{C}_s, \mu^*\mathcal{O}(k))$. Let $\ell > 0$ such that $\mathcal{I}_X(\ell)$ is generated by global sections. With $M = \overline{M}_{0,n}(\mathbb{P}^r, d, \bar{t})$ in

$$H^0\mathcal{I}_X(\ell)_M$$
$$\downarrow \qquad \searrow^{\rho}$$
$$H^0\mathcal{O}_{\mathbb{P}^r}(\ell)_M \longrightarrow \pi_*\mu^*\mathcal{O}(\ell)$$

we identify $\rho(F)$ over $s \in S$ with $\mu^*F_{|\mathcal{C}_s}$. Then $\mu(\mathcal{C}_s) \subset X \iff s \in Z = $ the zero scheme of $\{\rho(F) : F \in H^0\mathcal{I}_X(\ell)\}$.

Let then $\overline{M}_{0,n}(X, \beta, \bar{t}) = \{s \in Z : [\mu(\mathcal{C}_s)] = \beta\}$ be the component of Z determined by β. The group G acts on $\overline{M}_{0,n}(X, \beta, \bar{t})$; we construct $\overline{M}_{0,n}(X, \beta)$ by gluing the various quotients. $\square$

Note: this works also for general genus.

B) Assume X is nonsingular and convex. Recall that X convex
$\iff \forall \varphi : \mathbb{P}^1 \to X$, $H^1(\mathbb{P}^1, \varphi^*TX) = 0$, that is
$\iff \forall \varphi : \mathbb{P}^1 \to X$, $\varphi^*TX = \oplus\mathcal{O}(n_i)$, $n_i \geq 0$.

Claim: this is equivalent to $\forall \varphi : C \to X$, C at worst nodal, of arithmetic genus 0, $H^1(C, \varphi^*TX) = 0$.

The proof is by induction on the number of components. If $C = C' \cup L$, and $C' \cap L = \{p\}$, we have the exact sequence on C

$$0 \to \varphi^*TX \to \varphi^*TX_{|C'} \oplus \varphi^*TX_{|L} \to \varphi^*TX_p \to 0 \quad ,$$

whose long exact sequence yields the vanishing of $H^1(\varphi^*TX)$.

C) Local study of $M = \overline{M}_{0,n}(X,\beta,\overline{t})$. Let $[\mu] \in M$, that is, $[\mu]$ is an object $(\mu : C \to X; p_i, q_{ij})$. For $D = \operatorname{Spec} k[\epsilon]/(\epsilon^2)$, and $d_0 =$ the closed point of D,

$$TM_{[\mu]} = \{\varphi : D \to M | \varphi(d_0) = [\mu]\}$$
$$= \text{space of first order deformations of } (\mu : C \to X; p_i, q_{ij})$$
$$= \text{space of first order deformations of } (\mu : C \to X; p_i) =: \operatorname{Def}(\mu)$$

Let $\operatorname{Def}_G(\mu)$ be the space of first order deformations of $(\mu : C \to X; p_i)$, preserving the combinatorial type G of C. Then $\operatorname{Def}_G(\mu) \subset \operatorname{Def}(\mu)$, and its codimension is $\leq q =$ the number of nodes of C. We will show that this is in fact an equality.

Consider $\operatorname{Hom}(C,X) \times (C^n \setminus \text{diag.})$, containing the open subset $H^{st}_\beta(C)$ defined by $\{\text{stable } (\varphi; p_1, \dots, p_n) : [\operatorname{im} \varphi] = \beta\}$, mapping to $\overline{M}_{0,n}(X,\beta)$. Act on $H^{st}_\beta(C)$ with $\operatorname{Aut}(C)$:

$$\operatorname{Aut}(C) \times H^{st}_\beta(C) \to H^{st}_\beta(C)$$
$$(\psi, (\varphi; p_1, \dots, p_n)) \mapsto (\varphi \circ \psi; \psi(p_1), \dots, \psi(p_n))$$

This action is not free, but has finite stabilizers, giving étale maps $\operatorname{Aut}(C) \to$ orbit of $(\varphi; p_1, \dots, p_n)$.

Relativize this construction: for a flat family $\mathcal{C} \to S$ of $g = 0$ nodal curves, consider $\mathcal{C}^n = \underbrace{\mathcal{C} \times_S \cdots \times_S \mathcal{C}}_{n}$, $\operatorname{Hom}_S(\mathcal{C}, X \times S) \times_S (\mathcal{C}^n \setminus \text{diag})$ and its open subset $H^{st}_\beta(\mathcal{C})$, mapping (over S) to $\overline{M}_{0,n}(X,\beta,\overline{t})$. $\operatorname{Aut}_S(\mathcal{C})$ acts fiberwise with finite stabilizer on $H^{st}_\beta(\mathcal{C})$.

Now $\operatorname{Hom}_S(\mathcal{C}, X \times S)$ is an open subset of the Hilbert scheme of graphs. This allows us to compute its tangent space and dimension: for $\mu : \mathcal{C}_s \to X$,

—$T\operatorname{Hom}(\mathcal{C}_s, X)_{[\mu]} = H^0(\mathcal{C}_s, \mu^*TX)$;

—for X nonsingular, the dimension of all components of $\operatorname{Hom}_S(\mathcal{C}, X \times S)$ at $[\mu]$ is at least the expected one, that is $\dim H^0\mu^*TX - \dim H^1\mu^*TX + \dim S$;

—by convexity, $H^1 = 0$. The dimension equals the expected one.

The fibers of $\operatorname{Hom}_S(\mathcal{C}, X \times S) \to S$ have dimension $H^0\mu^*TX$; if S is smooth, $\operatorname{Hom}_S(\mathcal{C}, X \times S) \to S$ is smooth at $[\mu]$. It follows that $H^{st}_\beta(\mathcal{C}) \to S$ is smooth at $[\mu]$, of relative dimension $\dim H^0\mu^*TX + n$ (*).

Again let $\operatorname{Def}_G(C)$ be the space of first order deformations of C preserving the combinatorial type G, and consider the universal base space B of deformations of C preserving G. The space B can be seen to be smooth as follows. Stabilize C by replacing unstable components with marked points to get a marked curve C^{st} in some $\overline{M}_{0,m}$. The locus of curves in $\overline{M}_{0,m}$ determined by the combinatorial type of C^{st} is smooth (see Belorousski's talk on $\overline{M}_{0,n}$). Then B is a ball around C^{st} in this locus. Taking $S = B$ in the above, and $\mathcal{U} \to B$ the universal curve, we obtain $H^{st}_\beta(\mathcal{U}) \to \overline{M}_{0,n}(X,\beta)$ over B. Consider the

diagram

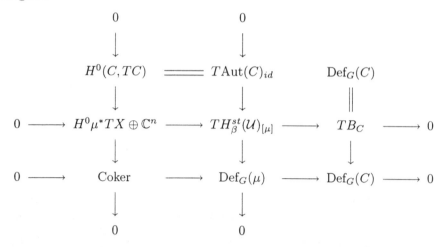

where the middle vertical map is the differential of $H_\beta^{st}(\mathcal{U}) \to \overline{M}_{0,n}(X,\beta)$. By Riemann-Roch and (*) we get:

$$\dim \mathrm{Def}_G(\mu) = \dim \mathrm{Def}_G(C) + \dim H^0\mu^*TX + n - \dim H^0(C,TC)$$

$$= \left(\sum_{|v|\geq 4} (|v| - 3) \right) + \left(\dim X + \int_\beta c_1 TX \right) + n - \left(\sum_{|v|\leq 3} (3 - |v|) \right)$$

where the $\sum$ is over the vertices v of G, and $|v|$ denotes valence. This gives

$$\dim \mathrm{Def}_G(\mu) = \dim X + \int_\beta c_1 TX + n - 3 - q$$

with $q = \#$ nodes of C.

Let $\mathcal{C} \to S$ be a smoothing of C, with S smooth. By (*) we have that $M_{0,n}(X,\beta)$ is dense in $\overline{M}_{0,n}(X,\beta)$, hence $\dim M_{0,n}(X,\beta) \leq \dim \mathrm{Def}(\mu)$. Since $H_\beta^{st}(\mathbb{P}^1) \to M_{0,n}(X,\beta)$ is surjective, $\dim M_{0,n}(X,\beta) = \dim X + \int_\beta c_1(TX) + n - 3$. Putting all together,

$$\dim X + \int_\beta c_1(TX) + n - 3 \leq \dim \mathrm{Def}(\mu) \leq \dim \mathrm{Def}_G(\mu) + q = \dim X + \int_\beta c_1(TX) + n - 3 \quad,$$

giving equalities through. Therefore

$$\dim T\overline{M}_{0,n}(X,\beta,\overline{t})_{[\mu]} = \dim \overline{M}_{0,n}(X,\beta,\overline{t})$$

and the space is smooth, as needed for Theorem 2.

D) Boundary. The boundary is studied by using the fact that $\overline{M}_{0,n}(\mathbb{P}^r, d, \overline{t})$ is a $(\mathbb{C})^r$-bundle over an open subset of $\overline{M}_{0,m}$ (where $m = n + d(r + 1)$). The conclusion is that the boundary of $\overline{M}_{0,n}(X,\beta,\overline{t})$ is a divisor with normal crossings.

9. The boundary of $\overline{M}_{0,n}(X,\beta)$—E. Rødland
November 19, 1996

Reminder of the construction: $\overline{M}_{0,n}(X,\beta)$ is obtained by gluing together quotients $\overline{M}_{0,n}(X,\beta,\bar{t})/G_{d,r}$, where $\overline{M}_{0,n}(X,\beta,\bar{t})$ are closed subschemes of schemes $\overline{M}_{0,n}(\mathbb{P}^r,d,\bar{t})$, in turn obtained as $(\mathbb{C}^*)^r$-covers of a Zariski-open $\mathcal{B}$ of $\overline{M}_{0,m}$, where $m = n + d(r+1)$. The boundary $\partial\overline{M}_{0,m}$ of $\overline{M}_{0,n}(X,\beta)$, that is the subset corresponding to maps from reducible curves, will be constructed from the boundary of $\overline{M}_{0,m}$.

Recall that $\partial\overline{M}_{0,m}$ is a normal-crossing divisor, with components $D(A|B)$, where $A \cup B = \{1,\ldots,m\}$: points of $D(A|B)$ correspond to curves $C = C_A \cup C_B$ with $C_A \cap C_B =$ point, and C_A, C_B resp. containing points marked from resp. A, B.

As $\mathcal{B}$ is Zariski-open in $\overline{M}_{0,m}$, the intersections $D(A|B) \cap \mathcal{B}$ will also be divisors crossing normally, and so will the pull-backs to $\overline{M}_{0,n}(\mathbb{P}^r,d,\bar{t})$. Restricting to $\overline{M}_{0,n}(X,\beta,\bar{t})$, the intersection is transversal enough so that the restricted boundary divisors still intersect with normal crossings (dimension count).

On the quotient $\overline{M}_{0,n}(X,\beta,\bar{t})/G_{d,r}$, the boundary will be a normal crossing divisor up to a finite group.

Let us describe the components of the boundary.

For $n = 0$: $\partial\overline{M}_{0,0}(X,\beta) = \cup_{\beta=\beta_A+\beta_B}D(\beta_A,\beta_B)$, where β_A, β_B are effective, and $D(\beta_A,\beta_B)$ consists of maps μ from $C = C_A \cup C_B$ with $C_A \cap C_B =$ point, and $\mu|_{C_A}, \mu|_{C_B}$ resp. represent β_A, β_B.

For $n > 0$, the situation is slightly more complicated. $\partial\overline{M}_{0,0}(X,\beta)$ is the union of components $D(A,B,\beta_A,\beta_B)$ where $A\cup B = \{1,\ldots,n\}$ and $A\cap B = \emptyset$; β_A, β_B are effective and adding up to β; if $\beta_A = 0$ then $|A| \geq 2$ and similarly for B (stability condition); and $D(A,B,\beta_A,\beta_B)$ consists of maps $C = C_A\cup C_B \xrightarrow{\mu} X$ where

a) $C_A \cap C_B =$ point, C_A, C_B genus-0 quasi-stable curves;
b) the markings from A (resp., B) are on C_A (resp., C_B);
c) $\mu_A = \mu|_{C_A}$, $\mu_B = \mu|_{C_B}$ represent β_A, β_B.

By the dimension computations from the results on deformations, one sees that the set of curves with $C_A \cong C_B \cong \mathbb{P}^1$ is dense in $D(A,B,\beta_A,\beta_B)$.

Denote by K the divisor $D(A,B,\beta_A,\beta_B)$, and by M_A the space $\overline{M}_{0,A\cup\{\bullet\}}(X,\beta)$, mapping to X by $\rho_A(C) =$ image of $\bullet$ (and similarly $M_B =$ etc.). Then we claim that K is 'almost' the product $M_A \times M_B$.

To be precise, consider

$$\widetilde{K} = M_A \times_X M_B = (\rho_A \times \rho_B)^{-1}(\Delta_X) \quad.$$

Letting similarly

$$\widetilde{K}(X,\bar{t}_A,\bar{t}_B) = M_A(X,\bar{t}_A) \times_X M_B(X,\bar{t}_B)$$

with the evident groups G_A, G_B acting on the factors, the map $\widetilde{K}(X, \bar{t}_A, \bar{t}_B)/$ $G_A \times G_B \to K$ induces $\psi : \widetilde{K} \to K$. If $C_A = \mathbb{P}^1$, $[\mu_A] \in \overline{M}_A(X, \bar{t}_A)$, we have

$$T_{M_A}([\mu_A]) \cong \mathrm{Def}(\mu_A) \twoheadrightarrow T_{M_{\{\bullet\}}}([\mu_A]) \cong H^0(\mu_A^* T_X / T_C(-p_0)) \to T_X(\mu_A(p_0))$$

Because $\mu_A^* T_X$ is generated by global sections, the second map too is surjective. Hence the composition, that is the differential of ρ_A (resp., ρ_B), is surjective, making $\widetilde{K}(X, \bar{t}_A, \bar{t}_B)$ smooth. It follows that $\widetilde{K}$ is locally normal, with finite quotient singularities.

To understand ψ, consider $A, B \neq \emptyset$ and $[\mu] \in K$, corresponding to a reducible curve $C = \cup C_i$. For $q_A \in A$ and $q_B \in B$, there is a unique path of components of C from q_A to q_B (since the components of C form a tree). Find this path $\{C_i\}_{i=1,\ldots,e}$ and order the components so that $q_A \in C_1$, $q_B \in C_e$. Denote by x_i the intersection $C_i \cap C_{i+1}$; and let $C_{A,i}$ be the closure of the connected component of $C \setminus \{x_i\}$ containing q_A (and define $C_{B,i}$ similarly). This yields a sequence of splittings of C into two components. For some i we will have that $\mu_*[C_{A,i}] = \beta_A$, and necessarily $\mu_*[C_{B,i}] = \beta_B$. Take the smallest such i. (Note: if $C_{A,j}$, $C_{A,j+1}$ realize the same class, then there must be at least one extra marking on C_j, by stability.)

This shows how to decompose $C = C_A \cup C_B$ uniquely, and in short that the map $\psi : \widetilde{K} \to K$ must be a bijection. Moreover K normal, ψ bijection $\implies \psi$ isomorphism, which is what we claimed.

For $A \neq \emptyset$ or $B \neq \emptyset$, or $\beta_A \neq \beta_B$ a similar discussion yields that ψ is bijective almost everywhere, hence birational.

For $A = B = \emptyset$ (so $n = 0$) and $\beta_A = \beta_B$, the set of maps from curves with two components is still dense in the corresponding $D(\ldots)$, but swapping the components will make ψ generically $2 : 1$ onto its image.

Part II

Topics in Quantum Cohomology

---

1. $QH^*(flag)$—W. Fulton
September 5, 1996

Two different things can be called the *quantum cohomology* of a flag manifold: the large ring, and the small ring.

Large: involves the numbers of maps $\mathbb{P}^1 \xrightarrow{f} X$, with prescribed $f_*[\mathbb{P}^1]$, that meet given general Schubert varieties $\Omega_1, \ldots, \Omega_t$;

Small: as above, but $t = 3$.

General story. Take a basis of the cohomology $H^*X = A^*X$, say $T_0 = 1$; $T_1, \ldots, T_p$ for A^1; and $T_{p+1}, \ldots, T_m$ for the rest. For X a flag manifold, the basis of classes of Schubert varieties is particularly effective: the product of two of these classes is a positive combination of these classes.

Define a power series:

$$\phi(y_0, \ldots, y_m) = \sum_{n_0 + \cdots + n_m \geq 3} \sum_{\beta \in A_1 X} I_\beta(T_0^{n_0} \cdots T_m^{n_m}) \cdot \prod \frac{y_i^{n_i}}{n_i!}$$

where $I_\beta(T_0^{n_0} \cdots T_m^{n_m})$ counts the number of maps f as above, touching n_i T_i's and with $\beta = f_*[\mathbb{P}^1]$. Note: T_0 and the divisor classes will 'factor out' easily in this definition, as curves meet divisors predictably.

Make $H^*X \otimes \mathbb{Q}[[y_0, \ldots, y_m]]$ into a $\mathbb{Q}[[y]]$-*algebra* by setting

$$T_i * T_j = \sum \phi_{ijk} g^{k\ell} T_\ell$$

where $\phi_{ijk} = \frac{\partial^3 \phi}{\partial y_i \partial y_j \partial y_k}$, and $g^{k\ell}$ is the inverse of the matrix $g_{k\ell}$ given by the ordinary intersection product on X: $g_{k\ell} = \int_X T_k \cdot T_\ell$. Note: for flag manifolds, the Schubert basis diagonalizes the intersection product, so this g is very simple.

THEOREM. *The product* $*$ *defined above is associative, with* T_0 *as unit element.*

The associativity of $*$ yields many relations between the I_β's: $\frac{m(m-1)(m^2-m+2)}{8}$, in fact. Often these relations alone and essentially trivial enumerative results suffice to determine all the I_β's.

EXAMPLE. For $X = F\ell(\mathbb{C}^6)$, $H^*X = \mathbb{Z}[x_1,\ldots,x_6]/(e_1,\ldots,e_6)$ (e_i: elementary symmetric polynomial of degree i); a $\mathbb{Z}$-basis for the cohomology is given by the *Schubert polynomials* S_w, $w \in S_6$. In principle one could write down the 35 billion (!) equations arising from the above, and derive the numbers; in practice, this is essentially undoable.

The *small* QH^* is defined similarly, but using

$$\overline{\phi}_{ijk}(y_0,\ldots,y_p) = \phi_{ijk}(y_0,\ldots,y_p,0,\ldots,0)$$

This $\overline{\phi}$ is easier to describe: $\overline{\phi}_{ijk}$ equals $\int_X T_iT_jT_k + \sum_{\beta\neq 0} I_\beta(T_iT_jT_k) \prod_{i=1}^p q_i^{\int_\beta T_i}$, with $q_i = e^{y_i}$.

The corresponding $*$ makes $QH^*X = H^*X \otimes \mathbb{Z}[q_1,\ldots,q_p]$ into a $\mathbb{Z}[q]$-algebra.

EXAMPLE. For $X = Gr(\ell,\mathbb{C}^n)$, $X = F\ell(\mathbb{C}^n)$, it is not too difficult to obtain $QH^*X = \mathbb{Z}[\ldots,q_1,\ldots,q_p]/(\text{explicit ideal})$. But note: this does *not* compute even the 3-point numbers $I_\beta(T_iT_jT_k)$: one needs "Quantum-Giambelli" formulas for the classes of T_i in QH^*.

More explicitly, take $X = Gr(\ell,\mathbb{C}^{n=k+\ell})$:

$$H^*X = \mathbb{Z}[\sigma_1,\ldots,\sigma_k]/(Y_{\ell+1},\ldots,Y_n)$$

with $\sigma_i = c_i(\text{quotient bundle})$, and $S_i = $ the determinant of the $i \times i$ matrix
$$\begin{pmatrix} \sigma_1 & \sigma_2 & \cdots & \sigma_i \\ 1 & \sigma_1 & \cdots & \\ & & \ddots & \\ & & & \sigma_1 \end{pmatrix}. \text{ Then:}$$

$$QH^*X = \mathbb{Z}[\sigma_1,\ldots,\sigma_k,q]/(S_{\ell+1},\ldots,S_{n-1}, S_n + (-1)^k q)$$

But again, this alone does not give the numbers! What one still needs is a quantum Giambelli. Surprisingly, this turns out to be the same recipe (with Young diagrams) as for the ordinary Giambelli: no 'quantum correction' is necessary.

Everything else is formal from here, and one can derive quantum-Pieri, quantum Littlewood-Richardson rules, etc.

For flag manifolds, $H^*F\ell(\mathbb{C}^n) = \mathbb{Z}[x_1,\ldots,x_n]/(e_1,\ldots,e_n)$ with $e_i = $ elementary symmetric polynomials. A basis consists of Schubert varieties $\mathfrak{S}_w$, with $w \in S_n$, corresponding to Schubert polynomials. So there must exist integer coefficients c_{uv}^w such that $\mathfrak{S}_u \cdot \mathfrak{S}_v = \sum c_{uv}^w \mathfrak{S}_w$; note: there is no known formula for the c_{uv}^w!

Fact: take any homogeneous presentation of $H^* X$, and find any q-deformation of the relations that hold in QH^*. Then the 'deformed' presentation computes QH^* (Fulton-Pandharipande).

For example, take the above presentation for $H^* X$, $X = F\ell(\mathbb{C}^n)$; deform the symmetric e_i to their quantum counterpart E_i; then

$$QH^* X = \mathbb{Z}[x_1, \ldots, x_n, q_1, \ldots, q_{n-1}]/(E_1, \ldots, E_n)$$

Here is an explicit description of the E_i. For the e_i, consider n marked dots:

$$x_1 \quad x_2 \quad x_3 \quad \cdots \quad x_n$$
$$\bullet \quad \bullet \quad \bullet \quad \cdots \quad \bullet$$

then the ordinary symmetric polynomials e_i are the sum of all monomials in the x_j's obtained by 'covering' i dots. For the E_i, one labels pairs of adjacent dots by q_i's:

$$x_1 \qquad x_2 \qquad x_3 \qquad \cdots \qquad x_n$$
$$\bullet \quad q_1 \quad \bullet \quad q_2 \quad \bullet \quad q_3 \quad \cdots \quad q_{n-1} \quad \bullet$$

and again one can write E_i as a sum of monomials corresponding to ways to 'cover' i dots; the q_i's (pairs of adjacent dots) must be disjoint, and will have degree 2 in the monomials. So for example $x_1 q_2 q_4$ will be a monomial in E_5 ($n \geq 5$).

There are quantum-Giambelli formula for quantum Schubert polynomials $\mathfrak{S}_w^q$ in $QH^* F\ell(\mathbb{C}^n)$, and quantum-Monk formulas. One remarkable thing about $\mathfrak{S}_w^q$: almost all (but not all as in the Grassmannian case) of them have no quantum correction.

---

2. The small quantum cohomology of the Grassmannian—R. Pandharipande
September 19, 1996

Reference: A. Bertram, [Bertram].

$G = G(k, n)$ will denote the Grassmannian of k-spaces in $\mathbb{C}^n$, with tautological sequence

$$0 \to S \to \mathbb{C}^n \to Q \to 0$$

First review the classical story.

(I) Additive structure of $H^*(G, \mathbb{Z})$. Fix a flag

$$0 = F_0 \subset F_1 \subset \cdots \subset F_n = \mathbb{C}^n$$

We have one *Schubert cell* for each non-increasing partition $\lambda = (\alpha_1, \ldots, \alpha_k)$: the corresponding closed Schubert cell is

$$\Omega_\lambda = \{V : \dim(V \cap F_{n-k+i-\alpha_i}) \geq i\}$$

and its class $[\Omega_\lambda]$ will be denoted by W_λ.

FACT. $\{W_\lambda\}$ is a $\mathbb{Z}$-basis of $H^*(G, \mathbb{Z})$.

(II) Multiplicative structure of $H^*(G, \mathbb{Z})$. Let σ_i be $c_i(Q)$, $1 \leq i \leq n - k$, and define formally polynomials $S_j(\sigma_1, \ldots, \sigma_{n-k})$ by

$$\frac{1}{1 + \sigma_1 t + \sigma_2 t^2 + \ldots} = \sum_{j=0}^{\infty} (-1)^j S_j t^j$$

Notice that $S_j(\sigma) = c_j(\check{S})$; so necessarily $S_j(\sigma) = 0$ for $k + 1 \leq j \leq n$ in $H^*(G)$: we get a set of relations between the σ's as elements of $H^*(G)$.

FACT. *This is a complete set of relations; that is,*

$$H^*(G, \mathbb{Z}) \cong \mathbb{Z}[\sigma_1, \ldots, \sigma_{n-k}]/(S_{k+1}, \ldots, S_n)$$

(III) Expressing W_λ as polynomials in the σ's. Giambelli's formula: with $\lambda = (\alpha_1, \ldots, \alpha_k)$ as before:

$$W_\lambda = |\sigma_{\alpha_i + i - j}| = \begin{vmatrix} \sigma_{\alpha_1} & \sigma_{\alpha_1+1} & \cdots & \\ \sigma_{\alpha_2-1} & \sigma_{\alpha_2} & & \\ \vdots & & \ddots & \\ \sigma_{\alpha_k-k+1} & & & \sigma_{\alpha_k} \end{vmatrix}$$

Also, since the W_λ's give an integral basis for $H^*(G)$, there must be integers $c_{\lambda\mu}^\gamma$ such that

$$W_\lambda \cdot W_\mu = \sum c_{\lambda\mu}^\gamma W_\gamma$$

There is in fact an explicit formula for the $c_{\lambda\mu}^\gamma$, given by the *Littlewood-Richardson rule*. A simpler case is for $W_{(m,0,\ldots,0)} = \sigma_m$, for which *Pieri's formula* says that

$$W_\lambda \cdot \sigma_m = \sum c_{\lambda m}^\gamma W_\gamma$$

with the $c_{\lambda m}^\gamma$ all zero, except if γ can be obtained from λ by adding m boxes to its Young diagram, but no two on the same column (see e.g. [Fulton], p. 264). In the latter, case the coefficient is 1.

Small quantum cohomology. Move now to the Quantum setting. First, define the 3-point Gromov-Witten invariants: the effective generator of $A_1 G(k, n) = \mathbb{Z}$ determines an identification of $A_1 G(k, n)$ with $\mathbb{Z}$; let

$$I_d(W_{\lambda_1}, W_{\lambda_2}, W_{\lambda_3}) = \begin{cases} \text{\# of rational curves in } G(k, n) \text{ of class } d, \\ \text{meeting general translates of representatives of} \\ \Omega_{\lambda_1}, \Omega_{\lambda_2}, \Omega_{\lambda_3} \text{ (if this number is finite; 0 otherwise)} \end{cases}$$

We will describe the quantum cohomology ring $QH^*(G) = QH^*(G(k, n))$ by retracing the classical steps.

(I) Additive structure. Additively, $QH^*(G(k,n))$ is the free $\mathbb{Z}[q]$-module $H^*(G) \otimes_{\mathbb{Z}} \mathbb{Z}[q]$. We have the obvious inclusion

$$H^*(G) \to QH^*(G)$$
$$W_\lambda \mapsto W_\lambda \otimes 1$$

Via this inclusion, the Schubert classes span a free basis of the quantum cohomology as a $\mathbb{Z}[q]$-module.

(II) Quantum product. There is a $*$-product which makes $QH^*(G)$ an associative, commutative $\mathbb{Z}[q]$-algebra with unit:

$$W_{\lambda_1} * W_{\lambda_2} = \sum_{d \geq 0} q^d \sum_\mu I_d(W_{\lambda_1}, W_{\lambda_2}, W_\mu) W_{\tilde\mu}$$

where $W_{\tilde\mu}$ is Poincaré dual to W_μ. For short, we will write $\langle W_{\lambda_1} W_{\lambda_2} W_\mu \rangle_d$ for $I_d(W_{\lambda_1}, W_{\lambda_2}, W_\mu)$.

REMARKS. (i) $\dim \overline{M}_{0,3}(G,d) = dn + \dim G$, so in order for a term in the sum not to be 0 it is necessary that

$$\operatorname{codim} W_{\lambda_1} + \operatorname{codim} W_{\lambda_2} + \operatorname{codim} W_\mu = dn + \dim G$$

In other words,

$$\operatorname{codim} W_{\lambda_1} + \operatorname{codim} W_{\lambda_2} = dn + \operatorname{codim} W_{\tilde\mu}$$

This shows that we can give a grading on $QH^*(G)$ compatible with the codimension grading on H^* and with $*$, if we take q to have degree n. $QH^*(G)$ is then a graded ring.

(ii) Since QH^* is graded, it is clear that only finitely many d's contribute to the $\sum_{d \geq 0}$ defining $*$.

(iii) $QH^*(G) \otimes (\mathbb{Z}[q]/(q)) \cong H^*(G)$. That is, the $d = 0$ contribution to $W_{\lambda_1} * W_{\lambda_2}$ is the usual intersection product $W_{\lambda_1} \cdot W_{\lambda_2}$ in the Grassmannian. $[G] = \sigma_0$ is the unit element for both $\cdot$ and $*$.

EXAMPLE. $G(1,n) = \mathbb{P}^{n-1}$. Here σ_1 is the class of a hyperplane, σ_{n-1} is the class of a point. What is $\sigma_1 * \sigma_{n-1}$?

—As point and hyperplane do not meet in $\mathbb{P}^n$, the contribution in degree 0 is 0;

—in degree 1: $\langle \sigma_1 \sigma_{n-1} \sigma_i \rangle_1$ is nonzero only for $k = n - 1$; for $i = n - 1$ it is the number of lines through two points and meeting a hyperplane, i.e., $\langle \sigma_1 \sigma_{n-1} \sigma_{n-1} \rangle_1 = 1$;

—in degree $d \geq 1$: $\langle \sigma_1 \sigma_{n-1} \sigma_i \rangle_d = 0$ by grading considerations, as q^d then has degree $dn > n - 1 + 1$.

Therefore $\sigma_1 * \sigma_{n-1} = q\sigma_0 = q$.

PROPOSITION. *The classes* $\sigma_1, \ldots, \sigma_{n-k} \in QH^*(G(k,n))$ *generate it as an algebra over* $\mathbb{Z}[q]$.

PROOF. Induction. It suffices to prove that $H^*(G) \otimes 1$ is contained in the subalgebra generated by $\sigma_1, \ldots, \sigma_{n-k}$. Consider then $\xi \in H^*(G)$; if codim $\xi = 0$, ξ is trivially in this subalgebra. For ξ of nonzero codimension ℓ, write $\xi = f(\sigma_1, \ldots, \sigma_{n-k})$ in H^* (that is, 'classically'); then we see that if computed in QH^* (with $*$ replacing $\cdots$)

$$(1) \qquad f(\sigma_1, \ldots, \sigma_{n-k}) = \xi + qA_1 + q^2 A_2 + \cdots + q^f A_f$$

with $A_i \in H^*(G)$ of lower codimension (by the grading). By induction the A_i's are in the subalgebra, and hence so is ξ by (1). $\square$

Hence we have a surjection from the polynomial ring

$$\mathbb{Z}[q, \sigma_1, \ldots, \sigma_{n-k}] \to QH^*(G)$$

and we seek generators for the kernel. Recall that the classical relations are the $S_{k+1}, \ldots, S_n$; S_j has codimension j.

REMARK. If codim W_{λ_1} + codim $W_{\lambda_2} \leq n-1$, then $W_{\lambda_1} * W_{\lambda_2} = W_{\lambda_1} \cdot W_{\lambda_2}$. This is again immediate from grading considerations, as $\deg q = n$. Therefore, the classical relations $S_{k+1}, \ldots, S_{n-1}$ must still hold in QH^*, with $*$ replacing

What happens to S_n?

The definition of the S_i's yields the formal polynomial identity

$$S_n(\sigma) - \sigma_1 S_{n-1}(\sigma) + \sigma_2 S_{n-2}(\sigma) + \cdots + (-1)^{n-k}\sigma_{n-k}S_k = 0$$

This holds in the polynomial ring, so it must hold in QH^*. However, we have just seen that $S_{k+1} = \cdots = S_{n-1} = 0$ in QH^*, therefore

$$S_n(\sigma) + (-1)^{n-k}\sigma_{n-k} * S_k = 0 \quad \text{in } QH^*(G).$$

Now,

σ_{n-k} = class of k-spaces containing a fixed line $\subset \mathbb{C}^n$;

$S_k(\sigma) = W_{\underbrace{(1, \ldots, 1)}_{k}}$ = class of k-spaces contained in a fixed $(n-1)$-space

$\subset \mathbb{C}^n$ (note that the classical S_k coincides with the quantum S_k by the above remark).

It is easy to conclude from this that the only nonzero contribution to $\sigma_{n-k} * S_k$ is

$$\langle \sigma_{n-k} \ W_{\underbrace{(1, \ldots, 1)}_{k}} W_{\underbrace{(1, \ldots, 1)}_{n-k}} \rangle_1 = 1$$

and the conclusion is that $\sigma_{n-k} * S_k = q$. Therefore, the relation involving $S_n(\sigma)$ in QH^* is

$$S_n(\sigma) + (-1)^{n-k}q = 0$$

Rank considerations show that these relations are all there is, and therefore

$$QH^*(G) = \mathbb{Z}[q, \sigma_1, \ldots, \sigma_{n-k}]/(S_{k+1}, \ldots, S_{n-1}, S_n + (-1)^{n-k}q)$$

This was first obtained by Witten, and Siebert-Tian.

(III). Surprisingly, Giambelli's formula holds in QH^*: that is, for every partition $\lambda = (\alpha_1, \ldots, \alpha_k)$

$$W_\lambda = \begin{vmatrix} \sigma_{\alpha_1} & \sigma_{\alpha_1+1} & \cdots \\ \sigma_{\alpha_2-1} & \sigma_{\alpha_2} & \\ \vdots & & \ddots \\ \sigma_{\alpha_k-k+1} & & \sigma_{\alpha_k} \end{vmatrix} \in QH^*(G(k,n))$$

where of course $*$ replaces $\cdot$ in computing the determinant. This is due to Bertram, and we will sketch a proof of this fact here.

First, we need an expression for products $W_{\lambda_1} * W_{\lambda_2} * \cdots * W_{\lambda_m}$. Define another Gromov-Witten-type invariant, by setting

$$\langle W_{\lambda_1} \ldots W_{\lambda_m} W_\mu \rangle_d = \begin{cases} \# \text{ of solutions (if finite) of the following} \\ \text{enumerative problem: fix general } p_1, \ldots, p_{m+1} \in \mathbb{P}^1, \\ \text{and general translates of } \Omega_\bullet, \text{ then count} \\ \text{the number of maps } f : \mathbb{P}^1 \to G \text{ with } [f(\mathbb{P}^1)] = d, \\ \text{and mapping } p_1 \text{ to } \Omega_{\lambda_1}, \ldots, p_m \text{ to } \Omega_{\lambda_m}, \\ \text{and } p_{m+1} \text{ to } \Omega_\mu. \end{cases}$$

(Note: these differ from the usual invariants in that we are fixing the $p_i \in \mathbb{P}^1$.)

PROPOSITION. $W_{\lambda_1} * \cdots * W_{\lambda_m} = \sum_{d \geq 0} q^d \sum_\mu \langle W_{\lambda_1} \ldots W_{\lambda_m} W_\mu \rangle_d W_{\tilde\mu}$.

PROOF. Let e_i $i = 1, \ldots, m+1$ be the evaluations maps from $\overline{M}_{0,m+1}(G, d)$ to $G(k,n)$, and let π be the forgetful map $\overline{M}_{0,m+1}(G, d) \to \overline{M}_{0,m+1}$. Rephrasing the definition of $\langle W_{\lambda_1} \ldots W_{\lambda_m} W_\mu \rangle_d$, we have

$$\pi_*(e_1^* W_{\lambda_1} \cdots e_{m+1}^* W_\mu) = \langle W_{\lambda_1} \ldots W_{\lambda_m} W_\mu \rangle_d [\overline{M}_{0,m+1}]$$

or

$$\langle W_{\lambda_1} \ldots W_{\lambda_m} W_\mu \rangle_d = e_1^* W_{\lambda_1} \cdots e_{m+1}^* W_\mu \cdot [\pi^{-1}(p)]$$

for arbitrary $p \in \overline{M}_{0,m+1}$.

CLAIM.

$$\langle W_{\lambda_1} \ldots W_{\lambda_m} W_\mu \rangle_d = \sum_{d_1+d_2=d} \sum_\nu \langle W_{\lambda_1} \ldots W_{\lambda_{m-1}} W_\nu \rangle_{d_1} \langle W_{\tilde\nu} W_{\lambda_m} W_\mu \rangle_{d_2}$$

This follows by choosing a general $p \in D(12 \ldots (m-1)|m(m+1))$.
The Claim implies the proposition, via an easy induction. $\square$

The 'Quantum-Giambelli' formula will follow in the end by applying Kempf-Laksov's formula, [K-L], which we recall here. Let M be a nonsingular variety, and consider a bundle map $\mathbb{C}^n \to E^{n-k}$ (note: not necessarily surjective).

Also, fix a flag $0 = F_0 \subset \cdots \subset F_n = \mathbb{C}^n$. Let D_{i,α_i} be the scheme-theoretic locus where $F_{n-k+i-\alpha_i} \to E$ has kernel of dimension $\geq i$. Finally, for $\lambda = (\alpha_1, \ldots, \alpha_k)$ let $C_\lambda = D_{1,\alpha_1} \cap \cdots \cap D_{k,\alpha_k}$. The formula states that if C_λ is pure of expected dimension, then $[C_\lambda]$ is given by the Giambelli determinant:

$$[C_\lambda] = \begin{vmatrix} \sigma_{\alpha_1} & \sigma_{\alpha_1+1} & \cdots & \\ \sigma_{\alpha_2-1} & \sigma_{\alpha_2} & & \\ \vdots & & \ddots & \\ \sigma_{\alpha_k-k+1} & & & \sigma_{\alpha_k} \end{vmatrix}$$

with $\sigma_j = c_j(E)$.

Also, we will need to use the *Quot scheme* (see for example [Strømme]) $Q_d = \mathrm{Quot}_{k,d}(\mathbb{C}^{n*}/\mathbb{P}^1)$ parametrizes (flatly) exact sequences

$$0 \to S \to \mathbb{C}^{n*} \to T \to 0$$

of quotient sheaves on $\mathbb{P}^1$, with $\mathrm{rk}(T) = k$, $\deg(T) = d$. There is a universal sequence of sheaves on $Q_d \times \mathbb{P}^1$:

$$0 \to \mathcal{S} \to \mathbb{C}^{n*} \to \mathcal{T} \to 0 \quad,$$

and we let $M_d \subset Q_d$ be the largest open subset such that the restriction of this universal sequence to $M_d \times \mathbb{P}^1$ is in fact a sequence of vector bundles. For $x \in M_d$ we have the exact sequence

$$0 \to \mathcal{S}_x \to \mathbb{C}^{n*}_x \to \mathcal{T}_x \to 0 \quad,$$

of vector bundles on $\mathbb{P}^1$; dualize:

$$0 \to \mathcal{T}_x^* \to \mathbb{C}^n_x \to \mathcal{S}_x^* \to 0$$

Thus we obtain from each $x \in M_d$ a rank-k subbundle of the trivial n-bundle over $\mathbb{P}^1$, and hence a degree-d map $\mathbb{P}^1 \to G(n,k)$; and conversely. In other words, we can think of Quot as a compactification of $M_{0,3}(G,d)$.

Quot is a nonsingular variety of dimension $dn + \dim G(k,n)$.

Note: in the above universal sequence

$$0 \to \mathcal{S} \to \mathbb{C}^{n*} \to \mathcal{T} \to 0 \quad,$$

$\mathcal{S}$ is in fact locally free (while $\mathcal{T}$ is not).

Now consider the map

$$\mathbb{C}^n \to \mathcal{S}^*$$

on $Q_d \times \mathbb{P}^1$. We will apply [K-L] to this map. First, define 'Schubert cycles' on Q_d. Note: as seen above, we have a map $M_d \times \mathbb{P}^1 \to G(k,n)$; morally we would like to choose $p \in \mathbb{P}^1$ and pull-back the usual Schubert cycles via

$$Q_d \times \{p\} \dashrightarrow M_d \times \{p\} \xrightarrow{e_p} G(k,n)$$

but we have to be careful as we go through the rational map. Schubert cycles can be defined in $M_d \times \{p\}$ by just setting $W_\lambda(p) = e_p^{-1}(\Omega_\lambda)$; define then

$$\overline{W}_\lambda(p) = \text{degeneracy locus in } Q_d \text{ of the corresponding bundle map } \mathbb{C}_p^n \to \mathcal{S}_p^*.$$

REMARKS. (i) For any points $p_1, \ldots, p_N \in \mathbb{P}^1$, and general translates of $\Omega_{\lambda_1}, \ldots, \Omega_{\lambda_N}$,

$$W_{\lambda_1}(p_1) \cap \cdots \cap W_{\lambda_N}(p_N) \subset M_d$$

is smooth of pure expected dimension, by Kleiman-Bertini.

(ii) If $p_1, \ldots, p_N$ are distinct points in $\mathbb{P}^1$, choose N general flags in $\mathbb{C}^n$ in defining the Ω_λ's; then

$$\overline{W}_{\lambda_1}(p_1) \cap \cdots \cap \overline{W}_{\lambda_N}(p_N) \subset Q_d$$

has pure expected dimension, and

$$W_{\lambda_1}(p_1) \cap \cdots \cap W_{\lambda_N}(p_N) \subset M_d$$

is Zariski dense in it. This is the main 'moving lemma' in Bertram's paper.

(iii) Therefore, we have an alternative definition for the new Gromov-Witten invariants:

$$\langle W_{\lambda_1} \ldots W_{\lambda_m} \rangle_d = \#(\overline{W}_{\lambda_1}(p_1) \cap \cdots \cap \overline{W}_{\lambda_N}(p_N))$$

if the latter is finite: indeed, both Quot and $\overline{M}_{0,3}(G,d)$ are compactifications of M_d; by (ii), $\cap W = \cap \overline{W}$ if finite; so we may measure the intersection number in $\overline{M}_{0,3}(G,d)$, which is the original definition of the invariants, by computing in Quot.

And now we get from [K-L] that (iv) $\overline{\sigma}_i(p) := c_i(\mathcal{S}_p^*)|_{Q_d}$ is independent of p; and that

$$\text{(v)} \qquad [\overline{W}_\lambda(p)] = \begin{vmatrix} \overline{\sigma}_{\alpha_1}(p) & \overline{\sigma}_{\alpha_1+1}(p) & \cdots & \\ \overline{\sigma}_{\alpha_2-1}(p) & \overline{\sigma}_{\alpha_2}(p) & & \\ \vdots & & \ddots & \\ \overline{\sigma}_{\alpha_k-k+1}(p) & & & \overline{\sigma}_{\alpha_k}(p) \end{vmatrix}.$$

Now we are ready to prove Quantum-Giambelli. Let $\Delta_\lambda(\sigma) \in QH^*(G)$ be Giambelli's determinant (with $*$-product). Extending $\langle \cdot \rangle_d$ by linearity to $\langle P(\sigma), W_\mu \rangle_d$ for all homogeneous polynomials P, we have

$$\Delta_\lambda(\sigma) = \sum_{d \geq 0} q^d \sum_\mu \langle \Delta_\lambda(\sigma), W_\mu \rangle_d W_{\check{\mu}}$$

(using the last proposition). Now by the classical Giambelli the $d = 0$ contribution in the sum equals W_λ; so it is enough to show that

$$\langle \Delta_\lambda(\sigma), W_\mu \rangle_d = 0 \quad \text{for } d \geq 1.$$

Computing on Q_d as in (iii):

$$\langle \Delta_\lambda(\sigma), W_\mu \rangle_d = \begin{vmatrix} \overline{\sigma}_{\alpha_1}(p_1) & \overline{\sigma}_{\alpha_1+1}(p_1) & \cdots \\ \overline{\sigma}_{\alpha_2-1}(p_2) & \overline{\sigma}_{\alpha_2}(p_2) & \\ \vdots & & \ddots \\ \overline{\sigma}_{\alpha_k-k+1}(p_k) & & \overline{\sigma}_{\alpha_k}(p_k) \end{vmatrix} \cdot \overline{W}_\mu(p_{k+1}) \quad ;$$

the $\overline{\sigma}_i(p)$ are independent of p by (iv), so we may choose $p_1 = \cdots = p_k$; by (v), the determinant then evaluates a cycle $\overline{W}_\lambda$:

$$\langle \Delta_\lambda(\sigma), W_\mu \rangle_d = \overline{W}_\lambda(p_1) \cdot \overline{W}_\mu(p_{k+1})$$
$$= \# \text{ maps } \mathbb{P}^1 \to G \text{ with } [f(\mathbb{P}^1)] = d,$$
$$\text{and } f(p_1) \in \Omega_\lambda, f(p_{k+1}) \in \Omega_\mu$$

The above intersection has pure expected dimension 0. However, if it is non-empty, it must have dimension at least 1 (since the automorphism of $\mathbb{P}^1$ with two markings acts). Hence, the intersection must be empty.

This proves that the $d \geq 1$ contributions vanish, and concludes the proof of the Quantum-Giambelli formula.

---

3. Rational curves on complete intersections in toric varieties (after Givental)—V. Batyrev
September 26, 1996

The starting point is the famous paper by Candelas, de la Ossa et al.[CDGP]: consider a hypersurface $V_5 \subset \mathbb{P}^4$ of degree 5; $c_1(V_5) = 0$ (that is, V_5 is a Calabi-Yau manifold). For each degree d, one expects a finite number n_d of rational curves of degree d (although this number was proved to be finite only for relatively small d). The mathematicians had shown that $n_1 = 2875$; $n_2 = 609250$, and they were in the process of computing n_3. [CDGP] claimed that they could compute the power series

$$K(q) = 5 + \sum_{d=1}^{\infty} n_d d^3 \frac{q^d}{1 - q^d}$$

How? Let $\phi_0(z) = \sum_{n=0}^{\infty} \frac{(5n)!}{(n!)^5} z^n$; and $\theta = z \frac{\partial}{\partial z}$; also, consider the operator

$$\mathcal{D} = \theta^4 - 5z(5\theta + 1)(5\theta + 2)(5\theta + 3)(5\theta + 4)$$

Then $\phi_0(z) = 0$ is the only regular solution at $z = 0$; there are three other solutions, with logarithmic singularities at $z = 0$. Set

$$\phi_0(z, \epsilon) = \sum_{n \geq 0} \frac{\Gamma(5(n + \epsilon) + 1)}{\Gamma(n + \epsilon + 1)^5} z^{n+\epsilon}$$

and differentiate formally with respect to ϵ:

$$\frac{\partial}{\partial \epsilon} \phi_0(z, \epsilon)|_{\epsilon=0} = \phi_1(z) = (\log z)\phi_0(z) + \psi(z) \quad , \quad \psi(o) = 0$$

Then

(*)
$$K(q) \left(\frac{dq}{q} \right)^{\otimes 3} = \frac{5}{(1 - 5^5 z)\phi_0^2(z)} \left(\frac{dz}{z} \right)^{\otimes 3}$$

with $q = \exp \frac{\phi_1(z)}{\phi_0(z)}$. This determines $K(q)$, giving a 'prediction' for the number n_d.

Later, Kontsevich computed n_4, showing it agrees with this prediction. Givental [Givental] found ways to go further, and managed to prove (*) rigorously. Note: it is very nontrivial that the n_d found this way should be nonnegative integers—the only known proof is via Givental's work.

Givental's framework covers many other cases. Denote by $V_{\ell_1,\dots,\ell_r} \subset \mathbb{P}^n$ a complete intersection of hypersurfaces of degrees $\ell_1,\dots,\ell_r$. Assume $\ell_1 + \cdots + \ell_r \leq n + 1$. Three cases are distinguished:

(1) $\ell_1 + \cdots + \ell_r \leq n - 1$
(2) $\ell_1 + \cdots + \ell_r = n$
(3) $\ell_1 + \cdots + \ell_r = n + 1$ (Calabi-Yau).

Denote by R the pull-back:

$$R : H^*(\mathbb{P}^n) \rightarrow H^*(V_{\ell_1,\dots,\ell_r}) \quad ;$$

for $H \in H^2(\mathbb{P}^n)$ the hyperplane class, $R(H)$ usually generates $\mathrm{Pic}X$.

In case (1): in the quantum cohomology ring $QH^*(V_{\ell_1,\dots,\ell_r})$, we have the following relation:
$$X^{n+1-r} = \ell_1^{\ell_1} \cdots \ell_r^{\ell_r} \cdot q \cdot X^{\sum \ell_i - r}$$

(so that the ring is graded, with $\deg q = n + 1 - \sum \ell_i$).

EXAMPLES. (i) $\mathbb{P}^k \subset \mathbb{P}^n$, $n = r - k$, $\ell_1 = \cdots = \ell_r = 1$, so this says $x^{k+1} = q$.
(ii) $G(2, 4) \subset \mathbb{P}^5$: get $X^5 - 2^2 q X = 0$.
(iii) $\mathbb{P}^1 \times \mathbb{P}^1 \subset \mathbb{P}^3$: consider X only, although here Pic$= \mathbb{Z} \oplus \mathbb{Z}$; then $X^3 - 4qX = 0$. (For X_1, X_2 generators of Pic, so that $X = X_1 + X_2$, $X_i^2 = q$.)

In case (2), one finds the slightly more complicated

$$\left(X + \prod_{i=1}^{r} \ell_i! q \right)^{n+1-r} = \prod_{i=1}^{r} \ell_i^{\ell_i} q \left(X + (\prod_{i=1}^{r} \ell_i!)q \right)^{n-r}$$

EXAMPLE. $\mathbb{P}^1 \hookrightarrow \mathbb{P}^2$, embedded as a conic; so $H \mapsto \mathcal{O}(2)$. Then this says

$$(X + 2q)^2 = 2^2 q(X + 2q)$$

that is: $X^2 = 4q^2$.

Givental's idea: define and use 'equivariant quantum cohomology'.

Reminder of usual equivariant cohomology. Given a topological space X and a Lie group G acting on X (typically $G \cong (S^1)^r$ or $(\mathbb{C}^*)^r$), define an equivariant cohomology ring $H_G^*(X)$ as follows: find a space EG which is contractible and on which G acts freely; set $BG = EG/G$; and define

$$H_G^*(X) = H^*((X \times EG)/G)$$

In particular, $H_G^*(pt) = H^*(BG)$.

EXAMPLE. $G = S^1$. Then $EG = S^\infty$ (the 'Hilbert sphere') $= \{(x_i) : \sum |x_i^2| = 1$, almost all $x_i = 0\}$. Then $H_G^*(pt) = H^*(EG/G) = H^*(\mathbb{CP}^\infty) = \mathbb{C}[x]$.

Properties of equivariant cohomology.

(1) A G-morphism $f : X_1 \to X_2$ induces a pull-back $f^* : H_G^*(X_2) \to H_G^*(X_1)$, a ring homomorphism, homogeneous of degree 0.

(2) A proper G-morphism $f : X_1 \to X_2$ induces a push-forward $f^* : H_G^*(X_1) \to H_G^*(X_2)$, homogeneous of degree $\dim X_2 - \dim X_1$.

In particular, the map $X \to pt$ defines an $H_G^*(pt) = H^*(BG)$-module structure on any $H_G^*(X)$; if X is proper, we also have a map $H_G^*(X) \to H_G^*(pt)$ (analogous to $\int$).

EXAMPLE. For $G = (S^1)^r$ (or $(\mathbb{C}^*)^r$), $H_G^*(pt) = \mathbb{C}[x_1, \ldots, x_r]$. The $\int$ of classes in $H_G^*(X)$ are polynomials in r variables for $G = S^1$.

EXAMPLE. $X = \mathbb{CP}^1$, $G = S^1$ or $\mathbb{C}$. Fix two points $0, \infty$, and corresponding inclusions i_0, i_∞ to $\mathbb{CP}^1$. Set $X_0 = i_{0*}[1]$, $X_\infty = i_{\infty*}[1] \in H_G^*(\mathbb{CP}^1)$; then $H_G^*(\mathbb{CP}^1) = \mathbb{C}[X_0, X_\infty]/(X_0 X_\infty)$. Here $H_G^*(pt) = \mathbb{C}[\hbar]$; the indeterminate $\hbar$ acts by multiplication by $X_0 - X_\infty$.

Back to Givental's work. Consider a map $\mathbb{P}^1 \to \mathbb{P}^n$, given by $(f_0 : \cdots : f_n)$ with f_i homogeneous in u, v, of degree d (so that the image is a degree-d curve). Also, let V_k be a hypersurface of degree k. For $n = 4$, a map as above is specified by $5(d+1)$ homogeneous coefficient, so we could take

$$L_d' = \mathbb{P}^{5(d+1)-1}$$

as a naive moduli space of mappings. Imposing that the image lies in a V_k for $k = 5$ amounts to $(5d + 1)$ conditions, giving a virtual dimension of 3 for the subscheme in $\mathbb{P}^{5(d+1)-1}$ of such rational curves. This accounts of course for the dim-3 group of automorphisms of $\mathbb{P}^1$; we can let $n_d =$ the length of the scheme defined by these conditions. For situation (1) ($\sum \ell_i \leq n - 1$), this naive method in fact works fine.

Main Lemma. For L_d = space of stable maps of degree $(d, 1)$ $C \to \mathbb{P}^4 \times \mathbb{P}^1$, there is a morphism $\mu : L_d \to L'_d$. For

$$C = C_0 \cup C_1 \cup \cdots \cup C_k \qquad C_i \cong \mathbb{P}^1$$

say C_0 has degree $(d', 1)$, C_i map to $\mathbb{P}^4 \times \{x_i\}$ and have degree $(d_i, 0)$ with $\sum d_i = d - d'$. $\mu(C)$ is given by $(g_0 f_0, \ldots, g_0 f_4)$, where (f_i) define $C_0 \to \mathbb{P}^4$ and $g_0 = \prod(x - x_i)$.

This can be done in every n. $\mathbb{P}^{(n+1)d-1} \times \mathbb{P}^1 \subset \mathbb{P}^{(n+1)(d+1)-1}$, where we think of the left-hand term as parametrizing degree-d rational curves, and the right-hand term as parametrizing degree-$(d-1)$ curves with a $\mathbb{P}^1$ tail. Imposing the curve to be on a degree-k V_k amounts to conditions bringing the dimension down to

$$1 + (n + 1)d - 1 - (k(d - 1) + 1)$$

If $k \leq n - 1$, this (the dimension of the moduli) is less than $(n + 1 - k)d_k - 1$; for $k = n$, the numbers will be the same and there will be a contribution from these curves.

Bott residue formula. Assume a compact X has an S^1(for example)-action with finitely many fixed points. Also, let E be an S^1-equivariant vector bundle, of rank $= \dim X = n$. Then

$$\int_X c_n(E) = \sum_{x \in X^{S^1}} \frac{\prod b_i}{\prod a_i}$$

where $a_i, b_i \in \mathbb{Z}$ are the weights of the S^1-action on $T_x X$ and E_x respectively, for $x \in X^{S^1}$.

In the context of enumerative geometry, this was first used in [E-S]. It motivated Kontsevich's work.

Now $(\mathbb{C}^*)^{n-1}$ acts on $\mathbb{P}^n$ (as the maximal torus of $\mathrm{PGL}(n + 1)$), and on L_d. The map $L_d \to L'_d$ is equivariant. Now we have to choose a good basis for the cohomology; unfortunately, there is no natural choice; but this can be done in equivariant cohomology. Recall that the pull-back gives maps $H_G^*(X) \to H_G^*(p)$ for all p in X, and for $p_i \in X^G$ in particular. So we have a map

$$H_G^*(X) \to \sum_i H_G^*(p_i)$$

THEOREM. *(Atiyah-Bott, "Localization theorem") This map is an isomorphism after inversion of some element in $H_G^*(pt)$.*

EXAMPLE. Two fixed points: $0, \infty$ for the action of S^1 on $\mathbb{C}\mathbb{P}^1$. This gives the obvious map

$$H_{S^1}^*(\mathbb{C}\mathbb{P}^1) = \mathbb{C}[X_0, X_\infty]/(X_0 X_\infty) \to \mathbb{C}[X_0] \oplus \mathbb{C}[X_\infty]$$

This is an isomorphism after inverting $\hbar = X_0 - X_\infty$.

This is the start of the rather tricky computation in Givental's paper.

Givental's method can be generalized to all Calabi-Yau complete intersections in toric varieties. Note: for Σ a fan, $\mathbb{P}\Sigma$ is not necessarily convex; for example, $\mathbb{P}^2$ blown-up at a point already is not.

There are predictions for the number of rational curves on Calabi-Yau complete intersection. For example, $V_{3,3} \subset \mathbb{P}^2 \times \mathbb{P}^2$: the $N_{0,d}$ have some periodic properties. It's not clear however what the physicists are counting.

Also, the relation between QH^* and Hodge theory is still mysterious.

Another possible generalization should be complete intersection in Grassmannians, for which the moduli space of stable maps should have all the necessary information (no mirror symmetry needed here).

Final comment: Givental does prove that $n_d \in \mathbb{Z}$; so far, however, it is only a 'virtual' number. To show that it is the actual number of rational curves may be very difficult; for example, this is not the case already for $V_{3,3} \subset \mathbb{P}^2 \times \mathbb{P}^2$.

---

4. Equivariant QH^* (after Givental)—B. Kim
October 3, 1996

Goal: to begin a study of the equivariant quantum cohomology ring QH_G^*, aiming to understand Givental's work.

First, recall the definition and basic properties of the classical equivariant cohomology ring H_G^*. Here

G is a connected compact Lie group;

X is an oriented smooth manifold or orbifold, acted upon by G;

X_G denotes the homotopic quotient, defined by $X \times_G EG$, where $EG \to BG$ is the universal G-bundle ($X \times_G EG = X \times EG/\sim$, where $(x, yg) \sim (gx, y)$ for $x \in X$, $y \in EG$, $g \in G$).

DEFINITION. $H_G^*(X) := H^*(X_G, \mathbb{C})$.

REMARK. $X \times_G EG \xrightarrow{\pi} BG$ is a fiber bundle, with fiber X, so π^* gives $H_G^*(X)$ an $H^*(BG)$-module structure. Suppose the Leray spectral sequence of $X_G \to BG$ degenerates at $E_2 = H^*(X) \otimes H^*(BG)$, so that $H_G^*(X) \cong H^*(X) \otimes H^*(BG)$ as $H^*(BG)$-modules (not as rings); and we have the exact sequence

$$0 \to I \cdot H_G^*(X) \to H_G^*(X) \to H^*(X) \to 0$$

with $I = \widetilde{H^*(BG)} = \ker(H^*(BG) \to H^*(pt))$ (the augmented cohomology group). (Note: the spectral sequence does degenerate in most algebro-geometric applications.) Then

(1) $H_G^*(X)$ is free over $H^*(BG)$; choose a basis $\{h_i\}$

(2) the composition $< \cdot, \cdot >: H_G^*(X) \otimes H_G^*(X) \xrightarrow{\cup} H_G^*(X) \xrightarrow{\pi_*} H^*(BG)$ is nondegenerate, and $\det(g_{ij} = < h_i, h_j >) \in \mathbb{C}$ (a priori $\in H^*(BG)$ only).

Equivariant Gromov-Witten invariants. Let X be a convex variety, acted upon by G. The evaluation/contraction diagrams

$$\overline{M}_{0,n}(X,d) \xrightarrow{ev_i} X$$
$$\downarrow \pi$$
$$\overline{M}_{0,n}$$

pass to the homotopic quotients:

$$\overline{M}_{0,n}(X,d)_G \xrightarrow{ev_i} X_G$$
$$\downarrow$$
$$\overline{M}_{0,n} \times BG$$

(note $\overline{M}_{0,n} \times BG = (\overline{M}_{0,n})_G$: the action of G is trivial here).
    Define $< \cdots >_{n,d} : H_G^{\otimes n}(X) \to H^*(BG)$ by

$$< \gamma_1 \cdots \gamma_n >_{n,d} = \int ev_1^*(\gamma_1) \cup \cdots \cup ev_n^*(\gamma_n) \in H^*(BG)$$

where $\int$ is the push-forward via $\overline{M}_{0,n}(X,d)_G \to BG$. Next, define the potential (up to quadratic terms)

$$\phi(\gamma) = \sum_{n,d} \frac{1}{n!} < \gamma^{\otimes n} >_{n,d} \in H^*(BG) \quad .$$

Choosing a basis $\{T_i\}$ of $H_G^*(X)$ over $H^*(BG)$, $\phi(\gamma^n)$ can be expanded as a formal power series:

$$(\phi_G =)\phi(\sum y_i T_i) = \sum(\cdots) \frac{y_1^{n_1} \cdots y_r^{n_r}}{n_1! \cdots n_r!}$$

(note: here $y_i \in H^*(BG)$.) For $x,y \in H^*(X_G)$, define $x \circ y$ by

$$< x \circ y, z >= \text{ partial derivative of } \phi \text{ in the directions } x, y, z$$

(which specifies it uniquely). One can prove that this $\circ$ is (super-)commutative, has a unit, is associative, and more.

    EXERCISE. Let $S^1 \times S^1$ act on $\mathbb{P}^1$ by $(z_1 : z_2) \mapsto (e^{2\pi is} z_1 : e^{2\pi it} z_2)$. Find ϕ_G of the quotient. (Answer: $QH_G^*(\mathbb{P}^1) = \mathbb{C}[u, v, \hbar_1, \hbar_2, q]/(u + v = \hbar_1 + \hbar_2, uv = \hbar_1 \hbar_2 + q)$.)

Applications. Computation of the small $QH^*(F)$, for F a partial flag manifold; and Givental's proof of the mirror conjecture in Calabi-Yau and Fano complete intersections in $\mathbb{P}^{n_1} \times \cdots \times \mathbb{P}^{n_r}$.
    For $\mathbb{CP}^n$ we can state a 'mirror theorem'. Consider a degree-ℓ hypersurface; the cases to be considered will be $\ell < n$, $\ell = n$, $\ell = n + 1$.

THEOREM (GIVENTAL). *The quantum differential equation on $\mathbb{CP}^n$ is hypergeometric.*

To clarify this statement, let $p = c_1(\mathcal{O}(1)) \in H^2(\mathbb{P}^n)$, and consider a $H^*(\mathbb{P}^n)$-valued formal function f:

$$f(t) = a_0(t)p^n + a_1(t)p^{n-1} + \cdots + a_n(t) \quad ;$$

here $e^t = q$, where q is the quantum correction in $QH^*(\mathbb{P}^n) = \mathbb{C}[p,q]/(p^{n+1}-q)$. The 'quantum differential equation' is

$$\frac{d}{dt}f(t) - p * f(t) = 0$$

where $*$ denotes the quantum product. Note that the equation implies in particular that $(\frac{d}{dt})^{n+1}a_0(t) = qa_0(t)$. The statement is that solutions to this differential equation have integral representations:

$$S(q) = \int_{\Gamma \subset Y_q} e^{u_0 + \cdots + u_n} \frac{du_0 \cdots du_n}{d(u_0 \cdots u_n)}$$

where $\omega = \frac{du_0 \cdots du_n}{d(u_0 \cdots u_n)}$ is the n-form defined by $du_0 \cdots du_n = \omega \wedge d(u_0 \cdots u_n)$ (so, $\omega = -\frac{1}{(n+1)q} \sum (-1)^i u_i du_0 \wedge \ldots \hat{du_i} \cdots \wedge du_n$). Also, $Y_q = \pi^{-1}(q)$ with $\pi : \mathbb{C}^{n+1} \to \mathbb{C}^*$, $(u_0, \ldots, u_n) \mapsto u_0 \cdots u_n$, and Γ is some real n-dimensional cycle in Y_q.

Choose coordinates in Y_q: $u_1, \ldots, u_n$, so that $u_0 = \frac{q}{u_1 \cdots u_n}$; $Y_q = \mathbb{C}^*_{u_1} \times \cdots \times \mathbb{C}^*_{u_n}$. Choosing $\Gamma = \prod$ unit circles, the integral above is

$$\int_{|u_i|=1, i=1, \ldots, n} e^{u_1 + \cdots + u_n + \frac{q}{u_1 \cdots u_n}} \frac{du_1 \cdots du_n}{u_1 \cdots u_n} = \sum \frac{q^d}{(d!)^{n+1}}$$

up to a normalization factor. This satisfies $(\frac{d}{dt})^{n+1} - q = 0$. The other linearly independent *multi-valued* solutions are obtained by different choices of Γ.

From the point of view of Morse theory: $g = -\text{Re}(F = u_0 + \cdots + u_n)$; $\Gamma = $ the unstable submanifold of ∇g in a Riemannian metric in Y_q. One can show that there are $(n+1)$ real n-dimensional unstable submanifolds Γ, and check that the corresponding $S(q)$ give a complete set of solutions for the quantum differential equation.

Next, consider a general quintic $X_q \subset \mathbb{P}^4$. The corresponding series is

$$\sum_d \frac{(5d)!}{(d!)^5} z^d \quad .$$

For a degree-ℓ hypersurface in $\mathbb{P}^N$, with $0 < \ell \leq N+1$, this would be $\sum \frac{(\ell d)!}{(d!)^{N+1}} z^d$. The above $S(q)$ should play the role of the $\ell = 0$-case of this expression.

Let $H \subset \mathbb{P}^N$ be a hypersurface of degree $\ell \leq N + 1$. Givental's proof of the mirror theorem for H amounts to the statement that the quantum differential equation, $D_1(U(q)) = 0$, will essentially coincide with the differential equation, $D_2(V(z)) = 0$, satisfied by $\sum \frac{(\ell d)!}{(d!)^{N+1}} z^d$. Here 'essentially' means *up to a change of coordinates* and more: for $\ell < N, z = q$; for $\ell = N$, $z^d = e^{-\ell! q} q^d$; for the Calabi-Yau case, $z = $ complicated.

A problem to overcome in this proof: while for $\mathbb{P}^n$ we knew the QH^*, we do not have as much for the generic hypersurface; hence, we cannot explicitly produce the quantum differential equation from the start. There are however complete solutions to the quantum differential equation in terms of intersections in $\overline{M}_{0,n}(X, d)$. The intersection theory so far was non-equivariant; equivariant intersection theory (equivariant under the action of the torus on $\mathbb{P}^n$) can be used to compute these solutions. So in a sense, although we do not have the quantum differential equation, we have its solutions. This reduces the computation to a suitable (and complicated) $\sum$ over trees. In fact, for $\ell < N$ the summation is simply a summation over chains, and yields recursion relations, from which the quantum differential equation can be recovered (and checked to agree with the non-Calabi-Yau mirror due to Givental's theory).

For $\ell = N$, $\sum$trees is a sum over chains, plus a correction; this correction can be evaluated with relative ease.

For $\ell = N + 1$, $\sum$trees is a sum over chains, plus several correction terms. Evaluating these is substantially harder.

---

5. QH^* of blow-ups of $\mathbb{P}^2$—L. Göttsche and R. Pandharipande
October 17, 1996

Part I (L. Göttsche). Let X_r denote the blow-up of $\mathbb{P}^2$ at r general points. The aim is to compute Gromov-Witten invariants of X_r, and show their enumerative significance in some cases.

Reminder on Gromov-Witten invariants. For X smooth and projective, and effective $\beta \in A_1 X$, there is a space $\overline{M}_{0,n}(X, \beta) = \{(\mu : C \to X; p_1, \ldots, p_n)\}$ with evaluation maps ρ_i to X, $i = 1, \ldots, n$. Then

$$I_\beta(\gamma_1 \cdots \gamma_n) = \int_{[\overline{M}_{0,n}(X,\beta)]} \rho_1^* \gamma_1 \cup \cdots \cup \rho_n^* \gamma_n$$

Note that we are not assuming that X is convex. $[\overline{M}_{0,n}(X, \beta)]$ is a natural fundamental class: the obvious one if X is convex; or a clever one otherwise, for example defined by means of the work of Behrend–Fantechi or Li–Tian.

More notations:

—the (pull-back of the) hyperplane class in X_r will be denoted H;

—the classes of the exceptional divisors will be E_1 through E_r;

—for $\alpha = (a_1, \ldots, a_r)$, (d, α) will be the divisor $dH - \sum a_i E_i$;

—$n_{d,\alpha}$ will be the expected dimension of $\overline{M}_{0,0}(X, (d, \alpha))$, that is $3d - 1 - \sum a_i$.

Write
$$N_{d,\alpha} = I_{(d,\alpha)}((\mathrm{pt})^{n_{d,\alpha}}) \quad :$$
since divisors factor out, these are the only 'interesting' Gromov-Witten numbers.

Intuitively, $N_{d,\alpha}$ is the number of rational curves in X_r of class (d,α) through $n_{d,\alpha}$ general points; that is, the number of rational curves in $\mathbb{P}^2$ of degree d, with points of multiplicity a_i at r given general points, and passing through $n_{d,\alpha}$ more general points.

These numbers ought to satisfy the following properties:

(P1) $N_{0,\alpha} = 0$ unless $\alpha = (0,0,\ldots,\overset{i}{-1},\ldots,0) =: -[i]$;

(P2) $N_{d,\alpha} = 0$ if $d > 0$ and any $a_i < 0$;

(P3) $N_{d,\alpha} = N_{d,\alpha_\sigma}$ for any permutation σ;

(P4) $N_{d,\alpha} = N_{d,(\alpha,0)}$;

(P5) If $n_{d,\alpha} > 0$, then $N_{d,\alpha} = N_{d,(\alpha,1)}$;

(P6) $N_{d,\alpha} = N_{d',\alpha'}$ if (d',α') is obtained from (d,α) by a Cremona transformation, that is, if

$$d' = 2d - a_1 - a_2 - a_3 \quad \text{and}$$

$$\alpha' = (d - a_2 - a_3, d - a_1 - a_3, d - a_1 - a_2, a_4, \ldots, a_r)$$

Further, one number is 'enumerative' (that is, it *does* count the appropriate number of rational curves) if so is the other.

PROOF OF (P6). Blow-up p_1, p_2 and p_3, obtaining exceptional divisors E_i, and proper transforms F_i of the lines L_i through them (L_1 through p_2 and p_3, etc.). Blow-down the F_i's to points q_i, and let $\overline{H}$ be the pull-back of the hyperplane from the blow-down. On the blow-up S there are two natural bases for Pic: $\{H, E_1, E_2, E_3\}$ and $\{\overline{H}, F_1, F_2, F_3\}$, with obvious relations

$$\overline{H} = 2H - E_1 - E_2 - E_3 \quad , \quad F_1 = H - E_2 - E_3 \quad , \quad \text{etc.}$$

For $x_4, \ldots, x_r$ additional general points on S, we may see the blow-up of S at $x_4, \ldots, x_r$ both as the blow-up of $\mathbb{P}^2$ at p_i, x_j and as the blow-up of (the other) $\mathbb{P}^2$ at q_i, x_j. From

$$dH - a_1 E_1 - \cdots - a_r E_r = (2d - a_1 - a_2 - a_3)\overline{H} - (d - a_2 - a_3)F_1 - \ldots$$

one gets $\overline{M}_{0,0}(X_r, (d,\alpha)) \cong \overline{M}_{0,0}(X_r, (d',\alpha'))$, from which (P6) follows. $\square$

Now the results are:

THEOREM 1. *The $N_{d,\alpha}$ are determined by simple recursion formulas.*

THEOREM 2. *The number of genus-0 stable maps with image (d,α) through $n_{d,\alpha}$ general points is finite. All of these maps are birational maps of $\mathbb{P}^1$ onto the image. Possibly counting multiplicities, $N_{d,\alpha}$ is this number.*

THEOREM 3. *Assume one of the following:*

(1) $n_{d,\alpha} > 0$;
(2) *There is an i for which $i \in \{1, 2\}$;*
(3) $r \le 8$.

Then all curves count with multiplicity 1, and each map is an immersion.

COROLLARY. *If $d \le 10$, then $N_{d,\alpha}$ is enumerative.*

Roughly speaking, the proof is in two parts. First, the associativity of the quantum product determines the numbers. Second, assume $n_{d,\alpha} = 0$; then the space $\overline{M}_{0,0}(X, (d, \alpha))$ has dimension 0 and consists only of curves as in Theorem 2. Under the conditions of Theorem 3, $\overline{M}_{0,0}(X, (d, \alpha))$ is *smooth* of dimension 0.

Digression on quantum cohomology for nonconvex varieties. Let X be a smooth projective variety, and $\mathcal{B} \subset H_2(X, \mathbb{Z})$ the cone of effective classes. Choose a $\mathbb{Z}$-basis $T_0, \underbrace{T_1, \ldots, T_p}_{\text{divisors}}, T_{p+1}, \ldots, T_m = \{pt\}$ for $H^*(X, \mathbb{Z})$, and let $\{T_i^\vee\}$ be the dual basis, so that $T_i \cdot T_j^\vee = \delta_{ij}$.

For variables $q_1, \ldots, q_p, y_{p+1}, \ldots, y_m$ set

$$\Gamma(q, y) = \sum_{n_{p+1} + \cdots + n_m \ge 0} \sum_{\beta \in \mathcal{B} - \{0\}} I_\beta(T_{p+1}^{n_{p+1}} \cdots T_m^{n_m}) q_1^{\int_\beta T_1} \cdots q_p^{\int_\beta T_p} \frac{y_{p+1}^{n_{p+1}} \cdots y_m^{n_m}}{n_{p+1}! \cdots n_m!}$$

Note: we are 'separating the β's', so that there is no question of convergence (β is reconstructed from $\int_\beta T_1, \ldots, \int_\beta T_p$ by duality).

Thinking $q_i = e^{y_i}$, set $\partial_i := \begin{cases} q_i \dfrac{\partial}{\partial q_i} & i = 1, \ldots, p \\[2mm] \dfrac{\partial}{\partial y_i} & i = p+1, \ldots, m \end{cases}$, and $\Gamma_{abc} = \partial_a \partial_b \partial_c \Gamma$.

We get a $\mathbb{Q}[[q, q^{-1}, y]]$-algebra structure on the free $\mathbb{Q}[[q, q^{-1}, y]]$-module generated by the T_i, by defining

$$T_i * T_j = (T_i \cdot T_j) + \sum_{r=1}^m \Gamma_{ijr} T_r^\vee$$

Fact: this is associative. This is shown by combining the properties of the virtual fundamental classes with the arguments for associativity from the convex case.

Back to X_r now. Here choose $T_0 = 1$, $T_1 = H$, $T_{i+1} = E_i$, $T_m = \{pt\}$. So

$$\Gamma(q, y) = \sum_{d, \alpha} N_{d,\alpha} q_1^d q_2^{a_1} \cdots q_{r+1}^{a_r} \frac{y_m^{n_{d,\alpha}}}{n_{d,\alpha}!}$$

$$T_i * T_j = (T_i \cdot T_j) T_m + \sum_{s=1}^m \epsilon_s \Gamma_{ijs} T_s + \Gamma_{ijm} T_0$$

with $\epsilon_s = 1$ if $s = 1$, -1 if $s > 1$.

LEMMA.

$$(g(m)) \qquad \Gamma_{mmm} = \sum_{s=1}^{m-1} \epsilon_s \left(\Gamma_{1sm}^2 - \Gamma_{11s}\Gamma_{smm} \right)$$

and, for $i = 2, \ldots, r+1 = m$,

$$(g(i-1)) \qquad \Gamma_{iim} - \Gamma_{11m} = \sum_{s=1}^{m-1} \epsilon_s \left(\Gamma_{1is}^2 - \Gamma_{11s}\Gamma_{iis} \right)$$

PROOF. For $g(m)$, look at the coefficient of T_0 in

$$(T_1 * T_1) * T_m - T_1 * (T_1 * T_m) \quad ;$$

for $g(i-1)$, look at the coefficient of T_1 in

$$(T_1 * T_1) * T_i - T_1 * (T_1 * T_i) \quad . \quad \square$$

Notation:

$$\vdash (d, \alpha) := \{((d_1, \beta), (d_2, \gamma)) : \text{both are} \neq 0;$$
$$\text{the sum is } (d, \alpha); \ n_{d_1,\beta} \geq 0; \ n_{d_2,\gamma} \geq 0; \ b_i \leq d_1, \ c_i \leq d_2\}$$

(where the b_i are the components of β, and the c_i are components of γ. With this, the basic recursion is given by

THEOREM. *The $N_{d,\alpha}$ are determined by*

$$N_{1,(0,\ldots,0)} = 1, N_{0,-[i]} = 1$$

and: if $n_{d,\alpha} \geq 3$,

$$N_{d,\alpha} = \sum_{\vdash (d,\alpha), d_i > 0} N_{d_1,\beta} N_{d_2,\gamma} \left(d_1 d_2 - \sum b_i c_i \right) \left(d_1 d_2 \binom{n_{d,\alpha} - 3}{n_{d_1,\beta} - 1} - d_1^2 \binom{n_{d,\alpha} - 3}{n_{d_1,\beta}} \right)$$

if $n_{d,\alpha} \geq 0$,

$$d^2 a_i^2 N_{d,\alpha} = (d^2 - (a_i^2 - 1)^2) N_{d,\alpha - [i]}$$
$$+ \sum_{\vdash (d,\alpha - [i]), d_i > 0} N_{d_1,\beta} N_{d_2,\gamma} \left(d_1 d_2 - \sum b_i c_i \right) (d_1 d_2 b_i c_i - d_1^2 c_i^2) \binom{n_{d,\alpha}}{n_{d_1,\beta}}$$

PROOF OUTLINE. $g(i)$ translates into something similar to $R(i)$; use this to prove (P1). After (P1) is proved, $g(i)$ does give $R(i)$, and $g(m)$ gives $R(m)$. (P2)—(P5) are proved by induction, using the recursions. $\square$

REMARK. Feeding the recursions into a computer, one can crunch out many examples. For all cases worked out so far, it so happens that two numbers are equal if and only if they must be equal by (P1)—(P6).

Part II (R. Pandharipande). Notations as in Göttsche's talk: we have $n_{d,\alpha} \geq 0$, $d > 0$; $\alpha = (\alpha_1, \ldots, \alpha_r)$, with $\alpha_i \geq 0$; X_r; etc. The enumerative geometry problem we consider is:

count the number of maps $[v] \in \overline{M}_{0,0}(X_r, (d, \alpha))$ *incident to* $n_{d,\alpha}$ *general points.*

In this lecture we address the following results:

(I) The number of solutions is finite; all are birational maps from $\mathbb{P}^1$ to the image in X_r; $N_{d,\alpha} \geq$ the number of solutions ≥ 0.

(II) Let at least one of the following two conditions hold:

(i) $n_{d,\alpha} > 0$;
(ii) $\alpha_i \in \{1, 2\}$ for some i.

Then $N_{d,\alpha}$ is the number of solutions. Moreover, they are immersions of $\mathbb{P}^1$ to X_r (questions: are they immersions to $\mathbb{P}^2$?)

Step 1. The trick we use is the "Elliptic herding" method (Harris–Caporaso, Kollár).

Let $(d, \alpha) \neq 0$, satisfying $n_{d,\alpha} < 0$. Then $\overline{M}_{0,0}(X_r, (d, \alpha)) = \emptyset$.

PROOF. Assume $d > 0$, and let $B_r = \underbrace{\mathbb{P}^2 \times \cdots \times \mathbb{P}^2}_{r} -$ diagonals. Consider the 'universal blow-up' $\mathcal{X}_r \xrightarrow{\pi} B_r$, whose fiber over $(b_1, \ldots, b_r)$ is the blow-up X_b of $\mathbb{P}^2$ at $b = (b_1, \ldots, b_r)$.

Let $\overline{M}_{0,0}(\mathcal{X}_r, (d, \alpha))$ be the π-relative space of maps. Let $\tau : \overline{M}_{0,0}(\pi, (d, \alpha)) \to B_r$ be the natural map. Assume τ is generically surjective. Since τ is proper, it must then be surjective. That is, specializing $(b_1, \ldots, b_r)$ we get stable curves in the limit.

Now 'herd' $b_1, \ldots, b_r$ to lie on a smooth elliptic curve: we get a stable map $\mu : C \to X_b$, with $b_i \in$ smooth elliptic curve $\mathcal{E} \subset \mathbb{P}^2$.

Simple numerology gives $C \cdot \mu^*(c_1(T_{X_b}) = 3d - \sum \alpha_i = n_{d,\alpha} + 1 \leq 0$.

Now the proper transform E of $\mathcal{E}$ is a section of T_{X_b} and the stable curve cannot have components on E. If $C = \cup C_i$, $C_i \cdot \mu^* E \geq 0$; so necessarily $C \cdot \mu^* E = 0$. Pushing forward, we get a stable curve which intersects $\mathcal{E}$ only along the blow-up points. This leads to a contradiction since the points are general. □

Step 2. Suppose (d, α) satisfies $n_{d,\alpha} \geq 0$. Every map $[\mu] \in \overline{M}_{0,0}(X_r, (d, \alpha))$ incident to $n_{d,\alpha}$ general points in X_r is a birational map $\mathbb{P}^1 \to X_r$; moreover, it is 'simply incident' to the points.

PROOF. This is easy using Step 1. We need to show that: the domain of μ is irreducible; μ is birational onto its image; and it is simply incident to the points. These are all shown by proving that failing any of these would contradict Step 1. For example, suppose that such a solution map has *reducible* source: $\mu : C = \cup C_i \to X_r$. There must be at least two components mapping nontrivially by μ (no marked points, so the tails must survive by stability). Let then $C_1(d_1, \alpha_1), \ldots, C_s(d_s, \alpha_s)$ be the components that are not collapsed, with $s \geq 2$. Note that $n_{d,\alpha} = s - 1 + \sum_{i=1}^s n_{d_i, \alpha_i} > \sum_{i=1}^s n_{d_i, \alpha_i}$. Also, C goes through the $n_{d,\alpha}$ points: let p_i =number of points contained in $\mu(C_i)$. Then $\sum_{i=1}^s p_i \geq n_{d,\alpha}$ (every point is in one of $C_1, \ldots, C_s$) $> \sum_{i=1}^s n_{d_i, \alpha_i}$. Then there must be a j such that $p_j > n_{d_j, \alpha_j}$; and this contradicts Step 1. Indeed, consider the map μ'_j from C_j to the blow-up of X_r at the p_j points in $\mu(C_j)$. The image has class $(d_j, (\alpha_j, m_1, \ldots, m_{p_j}))$; the virtual dimension would be

$$n_{d_j, \alpha_j} - \sum m_i \leq n_{d_j, \alpha_j} - p_j < 0$$

and the moduli space for this problem is empty by Step 1.

The other two parts to the argument are similar: the extra condition leads to negative virtual dimension. □

Step 3. Reduce to expected dimension 0. For this, blow-up the $n_{d,\alpha}$ (general) points producing a $X_{r+n_{d,\alpha}}$. The solutions to the original problem come from $\overline{M}_{0,0}(X_{r+n_{d,\alpha}}, (d, (\alpha, 1, \ldots, 1)))$. The recursions from Göttsche's section show that $N_{d,\alpha} = N_{d,(\alpha,1,\ldots,1)}$; so we may reduce to the case in which the expected dimension is 0 (note: the $X_{r+n_{d,\alpha}}$'s are even less convex than X_r!).

Step 4. Assume $n_{d,\alpha} = 0$. Then $\overline{M}_{0,0}(X_r, (d, \alpha))$ is pure of dimension 0 (not claiming here nonsingular or nonempty).

PROOF. By Step 2, we know $\overline{M}_{0,0}(X_r, (d, \alpha))$ equals the locus $\overline{M}_{0,0}^{\#}(X_r, (d, \alpha))$ of birational maps $\mathbb{P}^1 \to X_r$. Consider the normal sheaf sequence

$$0 \to T_{\mathbb{P}^1} \to \mu^* T_{X_r} \to N_{X_r|\mathbb{P}^1} \to 0$$

The Zariski tangent space to $[\mu]$ in $\overline{M}_{0,0}(X_r, (d, \alpha))$ is $H^0(\mathbb{P}^1, N_{X_r|\mathbb{P}^1})$; the degree of the normal bundle is $3d - \sum \alpha_i - 2 = n_{d,\alpha} - 1 < 0$.

This does not mean that $N_{X_r, \mathbb{P}^1}$ has no sections; but we have the torsion sequence

$$0 \to \text{Torsion} \to N_{X_r, \mathbb{P}^1} \to \text{Free} \to 0   :$$

the free part must have negative, and the torsion part positive, degree. The tangent space $H^0(\mathbb{P}^1, N_{X_r|\mathbb{P}^1})$ is equal to $H^0(\mathbb{P}^1, \text{torsion})$. It follows that the moduli space is of pure dimension 0. □

This concludes the proof of part I. We can only outline the proof to part II: if $n_{d,\alpha} = 0$, and $\alpha_i \in \{1, 2\}$ for some i, then $\overline{M}_{0,0}(X_r, (d, \alpha))$ is nonsingular, and each curve is immersed.

What is easy to see is that the nonsingularity of $\overline{M}_{0,0}(X_r, (d, \alpha))$ is equivalent to each curve being immersed. Indeed, refining the argument above, the degree of $N_{X_r, \mathbb{P}^1}$ must be -1. So $H^0(N_{X_r, \mathbb{P}^1}) = 0$ if and only if $N_{X_r, \mathbb{P}^1}$ is locally free, if and only if the curve is immersed.

By considering deformations of exceptional points $b_1, \ldots, b_r$, cohomological conditions on relevant normal sequences are obtained. These are enough to force the immersion condition. However, the relevant deformation argument works through only if there is at least one nonsingular or double point in the list, and this translates into the condition on α. □

6. Quantum Schubert polynomials—W. Fulton
November 7, 1996

This is in a sense a continuation of the first talk in this series, on the (small) quantum cohomology ring of flag varieties. The starting remark is that beyond an abstract description of QH^*, a *Giambelli formula* is needed to perform any computation, both in the classical and in the quantum case.

Review of the classical case. Notations: $X = F\ell(\mathbb{C}^n) = \{L_\bullet : L_1 \subset L_2 \subset \cdots \subset L_n = \mathbb{C}^n\}$ denotes a flag manifold; we have the universal flag of bundles

$$U_1 \subset U_2 \subset \cdots \subset U_n = V_X$$

over X, with $V = \mathbb{C}^n$; let $x_i = -c_1(U_i/U_{i-1})$; then

$$H^*X \cong \mathbb{Z}[x_1, \ldots, x_n]/(e_1^n, \ldots, e_n^n)$$

where $e_i^k = i$-th elementary symmetric polynomial in $x_1, \ldots, x_k$. The reason for this is that we have a map $\mathbb{Z}[x_1, \ldots, x_n]/(e_1^n, \ldots, e_n^n) \to H^*X$ since $e_i^n = (-1)^i c_i(V_X) = 0$ in H^*X; to see $\cong$, realizing X as a sequence of $\mathbb{P}^r$-bundles shows that $\mathrm{rk}\, H^*X = n!$ as needed.

Bases for H^*X over $\mathbb{Z}$:

(i) $x^I = x_1^{i_1} \cdots x_n^{i_n}$, with $i_j \leq n - j$ (so that $i_n = 0$ necessarily). This basis is preferred by algebraists.

(ii) $e_J = e_{j_1}^1 \cdots e_{j_{n-1}}^{n-1}$, with $0 \leq j_p \leq p$.

(iii) Classes of Schubert varieties: $[\Omega_w]$, with $w \in S_n$; to describe Ω_w, fix a flag $V_\bullet$ and set

$$\Omega_w = \{L_\bullet \,|\, \dim L_p \cap V_{n+1-q} \geq \#\{i \leq p \,|\, w(i) \leq q\} \forall p, q\}$$

Ω_w is an irreducible subvariety of codimension $= \ell(w) = \#\{i < j \,|\, w(i) > w(j)\}$. Of course this basis is preferred by geometers.

Giambelli problem: write $[\Omega_w]$ in $\mathbb{Z}[x]/(\ \ )$.

Solution: (Bernstein-Gelfand-Gelfand, Demazure, etc.)

(1) For $w = n\, n-1 \ldots 2\, 1$, $[\Omega_w] =$ the class of a point $= x_1^{n-1} x_2^{n-2} \cdots x_{n-1}^1$;

(2) Suppose $w(i) > w(i+1)$ for some i. Let $w' = w \cdot s_i$ (where $s_i = $ transposition $(i\, i+1)$), that is, interchange the values of i and $i+1$. Then

$$[\Omega_{w'}] = \partial_i [\Omega_w]$$

where ∂_i is a difference operator,

$$\partial_i P = \frac{P - s_i(P)}{x_i - x_{i+1}} \quad \text{(where } s_i(P) = P(\cdots, x_{i+1}, x_i, \cdots))$$

This recipe gives *polynomials*, denoted $\mathfrak{S}_w(x)$, representing $[\Omega_w]$. They have been defined, studied, and called *Schubert polynomials*, by Lascoux and Schützenberger.

REMARK. For any $u \in S_n$, we have an operator ∂_u: write $u = s_{i_1} \cdots s_{i_\ell}$ in the shortest possibly way; then $\partial_u = \partial_{i_1} \circ \cdots \circ \partial_{i_\ell}$. These operators are independent of the decomposition chosen for u, provided this has the shortest possible length. The Schubert polynomials $\mathfrak{S}_w$ can be written in terms of these operators.

EXAMPLES. $\mathfrak{S}_{12\ldots n} = 1$

$\mathfrak{S}_{s_i} = x_1 + \cdots + x_i$ $(\partial_i[\Omega_w] = 0$ if $w(i) < w(i+1)$; then $\mathfrak{S}_{s_i}$ cannot include any x_j for $j > i$, must be linear, etc.)

$n = 3$:

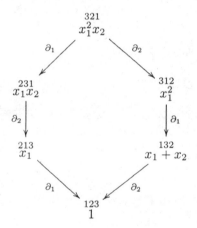

PROOF. (Of (2) from the Solution.) Let $X(i)$ = partial flag manifold (forget L_i). The natural map $X \to X(i)$ is a $\mathbb{P}^1$-bundle, $\mathbb{P}(U_{i+1}/U_{i-1})$ (abusing notations).

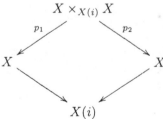

(1) $(p_1)_* \circ (p_2)^* : H^*X \to H^*X$ is ∂_i;

(2) p_1 maps $p_2^{-1}(\Omega_w)$ birationally onto $\Omega_{w'}$ (with notations as above).

This implies $[\Omega_{w'}] = \partial_i[\Omega_w]$, as needed. □

DIGRESSION ON SCHUBERT POLYNOMIALS. It is clear that $\mathfrak{S}_w = \sum a_I x^I$, with $a_I \in \mathbb{Z}$. In fact $a_I \geq 0$, but we do not know a 'geometric' reason for it. Kohnert (in his thesis) conjectured an intriguing formula for a_I. The formula is best illustrated in a simple example: take $w = 31524$; its *diagram* $D(w)$ is obtained like this:

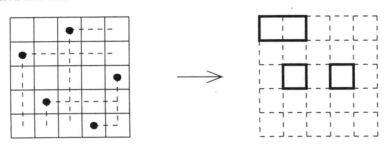

A *move* for such an arrangement consists of taking the box which is right-most in its row, and moving it up to the next available spot. Now play the following game: start with $D(w)$, and make all possible legal moves

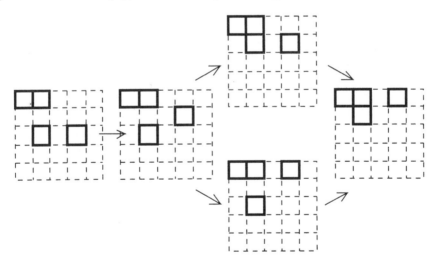

List all results D, each once, and let $x^D = \prod x_i^{\#\text{boxes in } i^{\text{th}} \text{ row of } D}$. Then the claim is that

$$\mathfrak{S}_w(x) = \sum x^D \quad .$$

In the example, this gives

$$\mathfrak{S}_{31524} = x_1^2 x_3^2 + x_1^2 x_2 x_3 + x_1^2 x_2^2 + x_1^3 x_3 + x_1^3 x_2$$

Although formulas for the a_I's have been proved, Kohnert's conjecture remains open.

(End of the digression)

Pieri problem: write the product of a general Schubert variety by a special one: $\mathfrak{S}_{s_i} \cdot \mathfrak{S}_w = ?$

Solution: (Monk, Chevalley) This is $\sum \mathfrak{S}_{w'} = \sum \mathfrak{S}_{wt_{ab}}$, where t_{ab} is the transposition $a \leftrightarrow b$; the $\sum$ is over $a \leq i < b$ such that $w(a) < w(b)$ and $w(j)$ is *not* between $w(a)$ and $w(b)$ for all j between a and b.

EXAMPLE. $\mathfrak{S}_{s_3} \cdot \mathfrak{S}_{31524} = \mathfrak{S}_{32514} + \mathfrak{S}_{41523}$
$\mathfrak{S}_{s_2} \cdot \mathfrak{S}_{31524} = \mathfrak{S}_{32514} + \mathfrak{S}_{51324} + \mathfrak{S}_{41523}$

REMARK. There must exist coefficients c_{uv}^w such that $\mathfrak{S}_u \cdot \mathfrak{S}_v = \sum c_{uv}^w \mathfrak{S}_w$. By geometry, $c_{uv}^w \in \mathbb{Z}_{\geq 0}$. No formula for all c_{uv}^w is even guessed!

Finally, the basis of Schubert varieties behaves well with respect to the intersection pairing:

$$\int_X [\Omega_u] \cdot [\Omega_{w_0 v}] = \delta_{uv} \quad ,$$

where $w_0 = n\, n-1 \ldots 2\, 1$.

Quantum version. Take variables $q_1, \ldots, q_{n-1}$, corresponding to $[\Omega_{s_1}], \ldots$
$\ldots, [\Omega_{s_{n-1}}]$; and duals $Y_i = [\Omega_{w_0 s_i}]$. Let $K = \mathbb{Z}[q_1, \ldots, q_{n-1}]$ ($\deg q_i = \int_{Y_i} c_1(TX) = 2$).

DEFINITION. $QH^* X = H^* X \otimes_{\mathbb{Z}} K$ as K-module; it is a K-algebra under

$$[V] * [W] = [V] \cup [W] + \sum_{d \neq 0} q^d I_d(V \cdot W \cdot \Omega_u)[\Omega_{w_0 u}]$$

with appropriate positions ($q^d = q_1^{d_1} \cdots q_{n-1}^{d_{n-1}}$, etc.)

Problem: Present QH^*: $QH^* X = K[x_1, \ldots, x_n]/(\ldots ? \ldots)$

For the relations, it suffices to find any deformations of $e_1^n, \ldots, e_n^n$ which hold in $QH^* X$.

THEOREM 1. *(Givental-Kim; Ciocan-Fontanine)*

$$QH^X = K[x_1, \ldots, x_n]/(E_1^n, \ldots, E_n^n)$$

where $E_i^k = i^{\text{th}}$ 'elementary quantum polynomial' in k variables.

These quantum polynomials are

$$E_i^k = \sum_{2|I|+|J|=i,\ I,J\ \text{'disjoint'}} q_I x_J \quad :$$

where q_i 'covers' $i, i+1$ and x_j covers j, and I, J are disjoint in the sense of covering disjoint subsets of $\{1, \ldots, k\}$. For example,

x_1		x_2		x_3		x_4		x_5		x_6		x_7		x_8		x_9
*	q_1	•	q_2	•	q_3	○	q_4	*	q_5	*	q_6	○	q_7	•	q_8	•

$q_2 q_8 x_1 x_5 x_6$ is a summand in E_7^9.

The original description of these polynomials was as follows: let A be the matrix

$$\begin{pmatrix} x_1 & q_1 & 0 & \cdots & & 0 \\ -1 & x_2 & q_2 & \cdots & & 0 \\ & & \ddots & & & \\ & & & \cdots & x_{k-1} & q_{k-1} \\ 0 & 0 & 0 & & -1 & x_k \end{pmatrix}$$

Then $\det(1 + \lambda A) = \sum E_i^k \lambda^i$. The two descriptions agree, as they both satisfy the recursion

$$E_i^k = E_i^{k-1} + x_k E_{i-1}^{k-1} + q_{k-1} E_{i-2}^{k-2} \quad .$$

Next, $c \mapsto c \otimes 1$ gives an inclusion

$$H^* X \hookrightarrow QH^* X = H^* X \otimes K$$

In the Grassmannian case, Giambelli's formula moves unchanged from the classical to the quantum ring (see Pandharipande's lecture on the Grassmannian).

THEOREM 2, 'QUANTUM GIAMBELLI'. *(Fomin-Gelfand-Postnikov, based on a result of Ciocan-Fontanine.) Write* $\mathfrak{S}_w = \sum n_{Jw} e_J$, *with* $e_J = e_{j_1}^1 \cdots e_{j_{n-1}}^{n-1}$ *and* $n_{jw} \in \mathbb{Z}$ *(note: the* n_{Jw} *are not necessarily positive). Then*

$$[\Omega_w] = \sum n_{Jw} E_J \, , \quad \text{with } E_J = E_{j_1}^1 \cdots E_{j_{n-1}}^{n-1}$$

THEOREM 3, 'QUANTUM MONK'. *(same people; also D. Petersen.) With* $\mathfrak{S}_w^q := \sum n_{Jw} E_J$,

$$\mathfrak{S}_{s_i}^q \cdot \mathfrak{S}_w^q = \underbrace{\sum \mathfrak{S}_{wt_{ab}}^q}_{\text{classical}} + \underbrace{\sum q_{cd} \mathfrak{S}_{wt_{cd}}^q}_{\text{quantum}} \quad ,$$

the second $\sum$ *over all* $c \leq i < d$ *such that* $w(c) > w(j) > w(d)$ *for all* j *between* c *and* d; *and where* $q_{cd} = q_c q_{c+1} \cdots q_{d-1}$.

PROOF. Here are four bases of $QH^* X$ over K:

   (i) x^I;
   (ii) E_J;
   (iii) $[\Omega_w]$;
   (iv) $\mathfrak{S}_w^q$.

We want to show that (iii) and (iv) are the same. Let K_+ be the cone spanned by all q^M, and let QH_+ be the cone spanned by all $q^M \cdot [\Omega_w]$, that is $\oplus K_+ \cdot [\Omega_w]$. Three facts:

   (a) QH_+ is closed under multiplication;
   (b) Each E_j^p is in QH_+, so each E_J is (this relies on Ciocan-Fontanine's result: $E_J^p = [\Omega_u]$ for $u = (p - j + 1, p - j + 2, \ldots, p + 1)$;

(c) This needs more notations: for $F \in QH^*X$, define $< F >$ by any of the following recipes:

expand in terms of basis (i), take coefficient of $x_1^{n-1} x_2^{n-2} \cdots x_{n-1}$; or

expand in terms of basis (iii), take coefficient of $[\Omega_{w_0}]$; or

expand in terms of basis (iv), take coefficient of $\mathfrak{S}_{w_0}^q$.

(It is easy to see that these coincide.) Remark: if $F \in QH_+$, then $< F > \in K_+$. The statement is then that

$$< \mathfrak{S}_u^q \cdot \mathfrak{S}_{w_0 v}^q >= \delta_{uv}$$

for all u, v. This is harder than it looks.

Also,

LEMMA. *Fix* $k \le \binom{n}{2}$. *For all* $w \in S_n$ *of length* $\ell(w) = k$, *there exist* $a_w > 0$, *with* $\sum_{\ell(w)=k} a_w \mathfrak{S}_w^q \in QH_+$.

Given (a), (b), (c) and the Lemma, we can prove Theorem 2:

Fix $w \in S_n$, let $k = \ell(w)$, and for $\ell \le \binom{n}{2}$ let $\ell' = \binom{n}{2} - \ell$.

(1) $[\Omega_w] = \mathfrak{S}_w^q + \sum_{\ell < k} P_{\ell w}$, with $P_{\ell w} = \sum_{\ell(v)=\ell} c_{vw} \mathfrak{S}_v^q$.

We have to show that all $P_{\ell w} = 0$, or equivalently that $< P_{\ell w} \cdot \mathfrak{S}_v^q >= 0$ for all v with $\ell(v) = \ell'$; or, equivalently, that

$$< P_{\ell w} \cdot E_J >= 0 \quad \text{for all } J \text{ with } |J| = \ell'.$$

(2) Now

$$< P_{\ell w} \cdot E_J >=< \sum \Omega_w \cdot E_j > \in K_+ \quad .$$

But $QH_+ \ni \sum_w a_w \mathfrak{S}_w^q = \sum a_w [\Omega_w] - \sum_{w,\ell} a_w P_{\ell w}$, so $- \sum_{w,\ell} a_w P_{\ell w} \in QH_+$, and

$$- \sum a_w < P_{\ell w} \cdot E_J > \in K_+ \quad .$$

It follows $< P_{\ell w} \cdot E_J >= 0$, as needed. □

REMARK. (c) can be proved in the form (c'): $< E_I \cdot E_J >= 0$ if $|I| + |J| \ge \binom{n}{2}$.

Comments on the algebra. Define operators $X_1, \ldots, X_n$ on $K[x_1, \ldots, x_n]$ by

$$X_k = x_k - \sum_{i<k} q_{ik} \partial_{t_{ik}} + \sum_{j>k} q_{kj} \partial_{t_{kj}}$$

(1) These operators commute, and commute with the $\cdot e_p^n$'s.

(2) $\forall f \in K[x]$, there is a unique $F \in K[X_1, \ldots, X_n]$ with $F(1) = f$.

(3) If $f = e_J$, then $F = E_J^*$; if $f = \mathfrak{S}_w(x)$, then $F = \mathfrak{S}_w^q(X)$.

The algebra of these operators streamlines the proofs considerably. For example, Quantum Monk amounts to proving

$$(X_1 + \cdots + X_i) \mathfrak{S}_w^q(X) = \underbrace{\sum \mathfrak{S}_{wt_{ab}}^q(X)}_{\text{classical}} + \underbrace{\sum q_{cd} \mathfrak{S}_{wt_{cd}}^q(X)}_{\text{quantum}} \quad :$$

and this is shown by evaluating at 1 and applying the above and the classical Monk.

EXAMPLE. Finally, here are the quantum Schubert polynomials for $n = 3$:

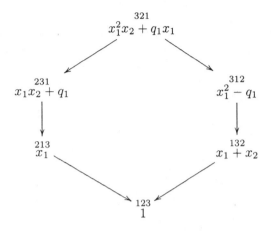

---

7. The small QH^*-ring of flag manifolds, I—I. Ciocan-Fontanine
November 14, 1996

Fix an n-dimensional complex vector space V. We will use the following notations (differing slightly from Fulton's notations):
$F = F\ell(V)$ will be the manifold of flags $\{U_1 \subset \cdots \subset U_{n-1} \subset V\}$; V_F will denote $V \otimes \mathcal{O}_F$; there are tautological bundles

$$E_1 \subset \cdots \subset E_{n-1} = E_n = V_F \twoheadrightarrow L_{n-1} \twoheadrightarrow \cdots \twoheadrightarrow L_1$$

on F, with $L_i = V_F/E_{n-i}$. We let x_i be $c_1(\ker(L_i \to L_{i-1}))$. Fact (see Fulton's lecture):
$$H^*(F) = \mathbb{Z}[x_1,\ldots,x_n]/(e_1^n,\ldots,e_n^n)$$

with e_i^k = the i-th symmetric polynomial in $x_1,\ldots,x_k$.
Next, fix a reference flag $V_1 \subset \cdots \subset V_{n-1} \subset V$. For $w \in S_n$ (=the symmetric group), let
$$r_w(q,p) = \#\{i \leq q : w(i) \leq p\} \quad \text{and set}$$
$$\Omega_w = \{y \in F / \operatorname{rk}_y(V_p \otimes \mathcal{O}_F \to L_q) \leq r_w(q,p), \forall q,p\}$$

The set Ω_w is irreducible, of ($\mathbb{C}$-)codimension $\ell(w)$ = length of w = the number of 'inversions' in w. We will denote by Ω_w also the corresponding elements in H_*F and H^*F. The set $\{\Omega_w : w \in S_n\}$ forms an additive basis for $H^*(F)$. Further, we have the duality

$$\int_F \Omega_w \Omega_{w_0 v} = \delta_{wv}$$

where w_0 is the permutation of maximal length, that is, $w_0(i) = n + 1 - i$.

The *Giambelli formula* tells us that $\Omega_w = \mathfrak{S}_w(x)$ in $H^*(F)$, with $\mathfrak{S}_w(x) =$ the corresponding Schubert polynomial, as defined in Fulton's lecture.

For $s_i = (i, i+1)$, the set $\{\Omega_{s_1}, \ldots, \Omega_{s_{n-1}}\}$ is a basis for $H^2(F)$. Aiming to quantize the situation, we let Y_i be $\Omega_{w_0 s_i}$, so that $\{Y_1, \ldots, Y_{n-1}\}$ gives a basis for $H_2(F)$, and let indeterminates q_i correspond to the Y_i. We consider the ring $K = \mathbb{Z}[q_1, \ldots, q_{n-1}]$ and define the (small) *quantum cohomology ring* of F to be

$$QH^*(F) = H^*(F) \otimes_{\mathbb{Z}} K$$

as a K-module, with product

$$\Omega_u * \Omega_v = \sum \overline{q}^{\overline{d}} I_{\overline{d}}(\Omega_u \Omega_v \Omega_w) \Omega_{w_0 w}$$

where $\overline{d} = (d_1, \ldots, d_{n-1})$, $\overline{q}^{\overline{d}} = \prod q_i^{d_i}$, etc. Note: F is a linear section of the Segre embedding of the product of the relevant Grassmannians in their Plücker embedding; Ω_{s_i} is the pull-back of $\mathcal{O}(1)$ from the ith Grassmannians. In particular, the Ω_{s_i} are nef, and it follows that if $f : C \to F$ is a map from a curve, then $f_*[C] = \sum d_i Y_i$, with all $d_i \geq 0$.

Next, we operate on polynomials $P(x, q)$ by replacing the intersection $\cdot$ with $*$: that is, we leave the q_i's alone, and we replace monomials in the x's with the corresponding quantum products. Note that as it turns out that $\Omega_{s_i} = \mathfrak{S}_{s_i}(x) = x_1 + \cdots + x_i = e_1^i$, we have $x_i = \Omega_{s_i} - \Omega_{s_{i-1}}$. So, for example, $P = q_4 x_1 x_2$ yields $\hat{P} = q_4(\Omega_{s_1} * (\Omega_{s_2} - \Omega_{s_1}))$. We want to write the Ω's in terms of this operation: that is,

the *Quantum Giambelli* problem is then to find $\mathfrak{S}_w^q(x, q)$ such that $\widehat{\mathfrak{S}_w^q} = \Omega_w$.

Necessarily, $\mathfrak{S}_w^q = \mathfrak{S}_w +$ a quantum correction; that is, $\mathfrak{S}_w^q(x, 0) = \mathfrak{S}_w$.

A presentation of QH^* is given by $\mathbb{Z}[x_1, \ldots, x_n, q_1, \ldots, q_{n-1}]/I_q$, where I_q is the ideal generated by 'quantum perturbations of the e_i^n'. It will follow from Theorem 1 below that these can be taken to be the polynomials $E_1^n, \ldots, E_n^n$:

$$E_i^k = i^{\text{th}} \text{ elementary quantum symmetric polynomial}$$

$$\text{in } x_1, \ldots, x_k, q_1, \ldots, q_{k-1}$$

(see also Fulton's lecture). These can be defined in terms of the characteristic polynomial of

$$\begin{pmatrix} x_1 & q_1 & 0 & \cdots & & 0 \\ -1 & x_2 & q_2 & \cdots & & 0 \\ & & \ddots & & & \\ & & & \cdots & x_{k-1} & q_{k-1} \\ 0 & 0 & 0 & -1 & x_k \end{pmatrix},$$

and satisfy the recursion

(*) $$E_i^k = E_i^{k-1} + x_k E_{i-1}^{k-1} + q_{k-1} E_{i-2}^{k-2}$$

Since $e_i^k = e_i^{k-1} + x_k e_{i-1}^{k-1}$, writing $E_i^k = e_i^k + {}^q E_i^k$, we have that (*) is equivalent to

$$(**) \qquad {}^q E_i^k = {}^q E_i^{k-1} + x_k {}^q E_{i-1}^{k-1} + q_{k-1} {}^q E_{i-2}^{k-2} \ .$$

Now denote by A_i^k the class represented by e_i^k in $H^*(F)$, $1 \le i \le k \le n$. We have

$$A_i^k = \begin{cases} \Omega_{\alpha_{i,k}} & k < n \\ 0 & k = n \end{cases}$$

with $\alpha_{i,k} =$ the cycle $(k-i+1, \ldots, k+1)$.

THEOREM 1. $\widehat{E_i^k} = A_i^k$ in $QH^*(F)$.

In particular, $\widehat{E_i^n} = 0$ in $QH^*(F)$; that is, $E_i^n \in I_q$ as promised.
Theorem 1 will follow from

THEOREM 2. *(Geometric formulation of a special case of the Quantum Monk formula.)*

$$\Omega_{s_j} * \Omega_{\alpha_{i,k}} = (classical\ term) + \delta_{jk} q_j \Omega_{\alpha_{i,k} \cdot s_j} = \begin{cases} classical & j \ne k \\ classical + q_k \Omega_{\alpha_{i-1,k-1}} & j = k \end{cases}$$

PROOF. (Of Theorem 1, assuming Theorem 2.) This is done by induction on k. The statement is trivial for $k = 1 \implies i = 1$: $e_1^1 = E_1^1 = x_1$. Assume proven for $\le (k-1)$; then

$$\widehat{e_i^k} = \widehat{e_i^{k-1}} + \widehat{x_k e_{i-1}^{k-1}} = \widehat{e_i^{k-1}} + x_k * \widehat{e_{i-1}^{k-1}}$$

$$= A_i^{k-1} - \widehat{{}^q E_i^{k-1}} + x_k * A_{i-1}^{k-1} - x_k * \widehat{{}^q E_{i-1}^{k-1}}$$

$$= A_i^{k-1} - \widehat{{}^q E_i^{k-1}} + x_k * A_{i-1}^{k-1} - x_k \widehat{{}^q E_{i-1}^{k-1}}$$

Now by Theorem 2

$$x_k * A_{i-1}^{k-1} = \Omega_{s_k} * \Omega_{\alpha_{i-1,k-1}} - \Omega_{s_{k-1}} * \Omega_{\alpha_{i-1,k-1}}$$

$$= A_i^k - A_i^{k-1} - q_{k-1} \Omega_{\alpha_{i-2,k-2}}$$

and therefore

$$\widehat{e_i^k} = A_i^k - \widehat{{}^q E_i^{k-1}} - q_{k-1} \widehat{E_{i-2}^{k-2}} - x_k \widehat{{}^q E_{i-1}^{k-1}}$$

$$= A_i^k - \widehat{B_i^k}$$

with $B_i^k = {}^q E_i^{k-1} + x_k {}^q E_{i-1}^{k-1} + q_{k-1} E_{i-2}^{k-2} = {}^q E_i^k$ by the recursion. That is,

$$\widehat{E_i^k} = \widehat{e_i^k} + \widehat{{}^q E_i^k} = A_i^k \ ,$$

completing the induction. $\square$

8. The small QH^*-ring of flag manifolds, II—I. Ciocan-Fontanine
November 21, 1996

Plan: as much as we can of the following

(0) What's left to prove;
(1) $\mathrm{Hom}_{\overline{d}}$ and its compactification $\mathcal{H}Q_{\overline{d}}$;
(2) Three-point (GW) numbers $I_{\overline{d}}(\Omega_{w_1}\Omega_{w_2}\Omega_{w_3})$ via $\mathcal{H}Q_{\overline{d}}$;
(3) Structure of the boundary;
(4) Proof of the moving lemma;
(5) Idea for computing $I_{\overline{d}}(\dots)$;
(6) Proof of 'special' Quantum-Monk formula;
(7) General Quantum-Monk formula.

(0) From the first lecture: $V \cong \mathbb{C}^n$, $F = F(V) =$ the space of complete flags in V.

Universal quotients: $V_F = V \otimes \mathcal{O}_F \to Q_{n-1} \to \cdots \to Q_1$;
Fixed reference flag: $V_1 \subset V_2 \subset \cdots \subset V_n = V$.
Every $w \in S_n$ determines a rank function,

$$r_w(q,p) = \#\{i : i \leq q; w(i) \leq p\}$$

The corresponding *Schubert variety* is $\Omega_w = \{\mathrm{rk}(V_p \otimes \mathcal{O} \to Q_q) \leq r_w(q,p) \forall q, p\}$; that is, the degeneracy locus of the universal quotients, using r_w for degeneracy conditions.

The codimension of Ω_w is the length of w, denoted $\ell(w)$.

If s_i denotes the transposition $(i, i+1)$, then $\{\Omega_{s_i}\}$ gives a (group) basis for $H^2(F)$.

The *quantum product* is defined by

$$\Omega_{w_1} * \Omega_{w_2} = \sum_{\overline{d}} \overline{q}^{\overline{d}} I_{\overline{d}}(\Omega_{w_1}\Omega_{w_2}\Omega_{w_3})\Omega_{w_0 w_3}$$

where $\overline{d} = (d_1, \ldots, d_{n-1})$, $\overline{q} = (q_1, \ldots, q_{n-1})$, and w_0 is the longest permutation. $\Omega_{w_0 w}$ is 'Poincaré dual' to Ω_w.

Then we have to show the following (*'Special Quantum-Monk'*):

$$\Omega_{s_j} * \Omega_{\alpha_{i,k}} = \begin{cases} \text{classical} & , \quad j \neq k \\ \text{classical} + q_k \Omega_{\alpha_{i-1,k-1}} & , \quad j = k \end{cases}$$

with $\alpha_{i,k} = s_{k-i+1} \cdots s_k$.

(1) Equivalent formulation:

$$I_{\overline{d}}(\Omega_{s_j}\Omega_{\alpha_{i,k}}\Omega_w) = \begin{cases} 1 & \overline{d} = \overline{e}_k, \text{ and } w = w_0\alpha_{i-1,k-1} \\ 0 & \text{otherwise} \end{cases}$$

Here $\bar{e}_k = (0, \ldots, \underset{k}{1}, 0, \ldots, 0)$.

The idea is to compute the three-point functions geometrically, that is as intersection numbers on some compactification of $\mathrm{Hom}_{\bar{d}}$.

We will denote $H_{\bar{d}} := \mathrm{Hom}_{\bar{d}}(\mathbb{P}^1, F) = M_{0,3}(F, \bar{d})$, that is

$$H_{\bar{d}} = \{[f] : f : \mathbb{P}^1 \to F, f_*[\mathbb{P}^1] = \sum d_i \Omega_{w_0 s_i}\}$$

We will use a compactification other than $\overline{M}_{0,3}(F, \bar{d})$.

Since F is homogeneous, $H_{\bar{d}}$ will be smooth of dimension $\binom{n}{2} + 2 \sum d_i$. We have the evaluation map $\mathbb{P}^1 \times H_{\bar{d}} \to F$, $(t, [f]) \mapsto f(t)$. For fixed distinct $t_1, t_2, t_3 \in \mathbb{P}^1$,

$$I_{\bar{d}}(\Omega_{w_1} \Omega_{w_2} \Omega_{w_3}) = \#\{[f] : f(t_i) \in \Omega_{w_i} \quad, \quad i = 1, 2, 3\} \quad .$$

We set $\Omega_w(t) := ev^{-1}(\Omega_w) \cap \{t\} \times H_{\bar{d}}$, so heuristically

$$I_{\bar{d}}(\Omega_{w_1} \Omega_{w_2} \Omega_{w_3}) = \#\{\Omega_{w_1}(t_1) \cap \Omega_{w_2}(t_1) \cap \Omega_{w_3}(t_1)\}$$

Explicitly,

$$\Omega_w(t) = \{\mathrm{rk}(V_p \otimes \mathcal{O} \to ev^* Q_q) \leq r_w(q, p) \forall p, q\} \cap \{t\} \times H_{\bar{d}}$$

Idea (Bertram): Compactify $H_{\bar{d}}$ so that there exist natural vector bundle extensions of $ev^* Q_q$ across the boundary, and use the same conditions to define $\Omega_w(t)$'s there.

The data of $\{f : \mathbb{P}^1 \to F, \bar{d}\}$ determines a sequence of surjections $\{V_{\mathbb{P}^1} \to L_{n-1} \to \cdots \to L_1\}$ with $\mathrm{rk}\, L_i = i$, $\deg L_i = d_i$. Dualize this sequence, and get

$$S_1 \subset S_2 \subset \cdots \subset S_{n-1} \subset V_{\mathbb{P}^1}^*$$

with $\mathrm{rk}\, S_i = i$, and $\deg S_i = -d_i$. And next consider the surjections

$$V_{\mathbb{P}^1}^* \to V_{\mathbb{P}^1}^*/S_1 \to \cdots \to V_{\mathbb{P}^1}^*/S_{n-1}$$

Degenerate this to get $\mathcal{H}Q_{\bar{d}}$, then set

$$\overline{\Omega}_w(t) := \{\mathrm{rk}(V_p \otimes \mathcal{O} \to \mathcal{S}_q^*) \leq r_w(q, p)\} \cap \{t\} \times \mathcal{H}Q_{\bar{d}} \quad .$$

The $\mathcal{S}_i$ are defined below:

$$\mathcal{S}_1 \subset \cdots \subset \mathcal{S}_{n-1} \to V^* \otimes \mathcal{O}_{\mathbb{P}^1 \times \mathcal{H}Q_{\bar{d}}}$$

THEOREM 1. *(Laumon, C-F, Kim) (i)* $\mathcal{H}Q_{\overline{d}}$ *is a smooth, irreducible projective variety of dimension* $\binom{n}{2} + 2\sum d_i$, *containing* $H_{\overline{d}}$ *as an open dense subscheme. There exists a universal sequence of quotients*

$$V^* \otimes \mathcal{O}_{\mathbb{P}^1 \times \mathcal{H}Q_{\overline{d}}} \to \mathcal{T}_{k-1} \to \cdots \to \mathcal{T}_1$$

with $\mathcal{T}_i$ *flat over* $\mathcal{H}Q_{\overline{d}}$ *and fixed relative Hilbert polynomial*

$$\chi(\mathcal{T}_i(m)) = (m+1)i + d_{n-i}$$

(This is an 'extension' of Grothendieck's quot-schemes).

(ii) $\mathcal{S}_i := \ker(V^* \to J_{n-i})$ *are vector bundles on* $\mathbb{P}^1 \times \mathcal{H}Q_{\overline{d}}$; *there are injections of sheaves*

$$\mathcal{S}_1 \subset \cdots \subset \mathcal{S}_{n-1} \subset V^*$$

(which may degenerate as maps of vector bundles).

(2) The following results show that we can use $\mathcal{H}Q_{\overline{d}}$ to define $I_{\overline{d}}(\Omega_{w_1}\Omega_{w_2}\Omega_{w_3})$:

THEOREM 2. *('Moving Lemma') (i)* $\forall w_1, \ldots, w_N \in S_n$, $t_1, \ldots, t_N \in \mathbb{P}^1$, *and general* $g_1, \ldots, g_N \in SL_n$, *the intersection* $\cap_{i=1}^N g_i\Omega_{w_i}(t_i)$ *is either empty, or of pure codimension* $\sum \ell(w_i)$ *in* $H_{\overline{d}}$.

(ii) If in addition the t_i's *are distinct, then* $\cap g_i\overline{\Omega}_{w_i}(t_i)$ *is either empty, or of codimension* $\sum \ell(w_i)$ *in* $\mathcal{H}Q_{\overline{d}}$, *and equals the closure of the* $\cap\Omega$ *in (i).*

In particular, when the codimension is maximal (that is, when the intersection consists of points) it all happens in the open part, so it counts what it is supposed to count.

COROLLARY 1. *The class of* $\overline{\Omega}_w(t)$ *in* $A^{\ell(w)}(\mathcal{H}Q_{\overline{d}})$ *is independent of* t *and the fixed flag* $V_\bullet \subset V$.

COROLLARY 2. *If* $\sum \ell(w_i) = \binom{n}{2} + 2\sum d_i$ *and* $t_1, \ldots, t_N$ *are distinct, then the number of intersection points of general translates is given by*

$$\# \cap_{i=1}^N \Omega_{w_i}(t_i) = \int_{\mathcal{H}Q_{\overline{d}}} [\overline{\Omega}_{w_1}(t_1)] \cup \cdots \cup [\overline{\Omega}_{w_N}(t_N)]$$

So we define

$$\langle \Omega_{w_1} \cdots \Omega_{w_N} \rangle_{\overline{d}} = \begin{cases} \int_{\mathcal{H}Q_{\overline{d}}} \cdots & \text{if } \sum \ell(w_i) = \binom{n}{2} + 2\sum d_i \\ 0 & \text{otherwise} \end{cases}$$

(Difference with the usual GW: here we take a fixed, albeit general, configuration of points in $\mathbb{P}^1$.)

COROLLARY 3. $\langle \Omega_{w_1} \Omega_{w_2} \rangle_{\overline{d}} = 0$ for $\overline{d} \neq 0$.

(Indeed, if we get one we must get a whole $\mathbb{C}^*$ of intersections.)

However, since ev does not extend to $\mathcal{H}Q_{\overline{d}}$, a different strategy is needed to prove (ii). We need to analyze the restriction of $\cap \overline{\Omega}_{w_i}(t_i)$ to the boundary, and show that it has 'large enough' codimension.

(3): Structure of the boundary. Recall the universal sequence on $\mathbb{P}^1 \times \mathcal{H}Q_{\overline{d}}$:

$$0 \to \mathcal{S}_1^{\overline{d}} \to \cdots \to \mathcal{S}_{n-1}^{\overline{d}} \to V_{\mathbb{P}^1 \times \mathcal{H}Q_{\overline{d}}}^* \to \mathcal{T}_{n-1}^{\overline{d}} \to \cdots \to \mathcal{T}_1^{\overline{d}}$$

$H_{\overline{d}}$ is the largest open subscheme in $\mathcal{H}Q_{\overline{d}}$ such that on $\mathbb{P}^1 \times H_{\overline{d}}$ all these maps are nondegenerate as vector bundle maps.

Next, we 'stratify' according to the degeneracies of the map, more precisely according to the ranks of $\mathcal{T}_i^{\overline{d}}$.

Idea: how to construct $V_{\mathbb{P}^1}^* \to T_{n-1} \to \cdots \to T_1$ on $\mathbb{P}^1$ with assigned Hilbert polynomials and with prescribed ranks at $t \in \mathbb{P}^1$? Let $\text{rk}_t\, T_{n-i} = n - i + e_i$, with $e_i \geq 0$; start with $S_1 \subset \cdots \subset S_{n-1} \subset V_{\mathbb{P}^1}^*$, a point in $\mathcal{H}Q_{\overline{d}-\overline{e}}$, together with quotients $S_i(t) \to \mathbb{C}^{e_i}(t)$. Let $\widetilde{S}_i = \ker(S_i \to \mathbb{C}^{e_i}(t))$, a vector bundle of rank i and degree $-d_i$ on $\mathbb{P}^1$, and a subsheaf of $V_{\mathbb{P}^1}^*$. If we want $\widetilde{S}_i \to V^*$ to factor through $\widetilde{S}_{i+1}$, we should start with quotients $S_i(t) \to \mathbb{C}^{e_i}(t)$ together with compatible maps $\mathbb{C}^{e_i}(t) \to \mathbb{C}^{e_{i+1}}(t)$. Set then $T_{n-i} = V^*/\widetilde{S}_i$.

The following construction and theorem are a globalization of this idea, showing also that every degeneration is obtained by the above construction.

Let $\overline{e} = (e_1, \ldots, e_{n-1})$ such that $0 \leq e_i \leq \min(i, d_i)$, and $e_i - e_{i-1} \leq 1$. Consider the universal sequence

$$0 \to \mathcal{S}_1^{\overline{d}-\overline{e}} \hookrightarrow \cdots \hookrightarrow \mathcal{S}_{n-1}^{\overline{d}-\overline{e}} \to V_{\mathbb{P}^1 \times \mathcal{H}Q_{\overline{d}-\overline{e}}}^*$$

Let $G_i \xrightarrow{\pi_i} \mathbb{P}^1 \times \mathcal{H}Q_{\overline{d}-\overline{e}}$ be the Grassmann bundle of e_i-dimensional quotients of $\mathcal{S}_i^{\overline{d}-\overline{e}}$, with universal sequence

$$0 \to K_i \to \pi_i^* \mathcal{S}_i^{\overline{d}-\overline{e}} \to Q_i \to 0$$

and set

$$X_{\overline{e}} = G_1 \times_{\mathbb{P}^1 \times \mathcal{H}Q_{\overline{d}-\overline{e}}} \cdots \times_{\mathbb{P}^1 \times \mathcal{H}Q_{\overline{d}-\overline{e}}} G_{n-1} \quad .$$

Define $\mathcal{U}_{\overline{e}} \subset X_{\overline{e}}$ as the subset where $K_i \to Q_{i+1}$ vanishes, and $K_i \to V^*$ is injective as a vector bundle map.

THEOREM 3. (i) $\mathcal{U}_{\overline{e}}$ is irreducible, Cohen-Macaulay, of dimension $1 + \binom{n}{2} + 2\sum d_i - \sum e_i - \sum e_i(e_i - e_{i-1})$. The projection $\pi : \mathcal{U}_{\overline{e}} \to \mathbb{P}^1 \times \mathcal{H}Q_{\overline{d}-\overline{e}}$ is flat, and its image contains $\mathbb{P}^1 \times H_{\overline{d}-\overline{e}}$.
(ii) There exist maps $h_{\overline{e}} : \mathcal{U}_{\overline{e}} \to \mathcal{H}Q_{\overline{d}}$ satisfying
(a) if $\text{rk}_{(t,x)}\, \mathcal{T}_{n-i}^{\overline{d}} = n - i + e_i$, then $x \in h_{\overline{e}}(\mathcal{U}_{\overline{e}})$;
(b) the restriction of $h_{\overline{e}}$ to $\pi^{-1}(\mathbb{P}^1 \times H_{\overline{d}-\overline{e}})$ is an isomorphism onto its image.

EXAMPLES. 1) $\overline{e}_i = (0, \ldots, 0, \underset{i}{1}, 0, \ldots, 0)$. Then X_{e_i} is a $\mathbb{P}^{i-1}$-bundle over $\mathbb{P}^1 \times \mathcal{H}Q_{\overline{d}-\overline{e}_i}$ and $\mathcal{U}_{\overline{e}_i}$ is a section over an open set, that is, $\pi : \mathcal{U}_{\overline{e}_i} \to \mathbb{P}^1 \times \mathcal{H}Q_{\overline{d}-\overline{e}_i}$ is an open immersion.

2) $n = 3$, $\overline{d} = (1,1)$. We have the strata D_1, D_2 from above and a codimension 2 stratum $E = h_{(1,1)}(\mathcal{U}_{(1,1)})$; $\mathcal{U}_{(1,1)} = X_{(1,1)} = \check{\mathbb{P}}(\mathcal{S}_2^{(0,0)})$; $h_{(1,1)}$ maps $\mathcal{U}_{(1,1)}$ isomorphically onto E.

LEMMA 1. $h_{\overline{e}}^{-1}(\overline{\Omega}_w(t)) = \pi^{-1}(\mathbb{P}^1 \times \overline{\Omega}_w(t)) \cup \widetilde{\Omega}_w^{\overline{e}}(t)$

where $\widetilde{\Omega}_w^{\overline{e}}(t) \subset \mathcal{U}_{\overline{e}}(t) := \pi^{-1}(\{t\} \times \mathcal{H}Q_{\overline{d}-\overline{e}})$ is defined by $\{\mathrm{rk}(V_p \otimes \mathcal{O} \to K_q^*) \leq r_w(q,p), \forall p, q\}$.

On $\mathcal{U}_{\overline{e}}(t)$ we have the flag of quotients

$$V \otimes = \mathcal{O} \twoheadrightarrow K_{n-1}^* \twoheadrightarrow K_{n-2}^* \twoheadrightarrow \cdots \twoheadrightarrow K_1^*$$

with $\mathrm{rk}\, K_i^* = i - e_i$, implying that some of these maps are isomorphisms. Let $k = \#\{i - e_i\}$. Define a partition of $[0, n]$ as follows: $i_0 = 0$, $i_j = \min\{i | i - e_i > i_{j-1} - e_{i_{j-1}}\}$, $i_{k+1} = n$. Let $n_j = i_j - e_{i_j}$, i.e., $i \in [i_j, i_{j+1} - 1] \implies \mathrm{rk}\, K_i^*$ is n_j (constant). Since the matrix $(r_w(q,p))$ has nondecreasing columns, $\widetilde{\Omega}_w^{\overline{e}}$ is defined by $\{\mathrm{rk}(V_p \otimes \mathcal{O}) \to K_j^* \leq r_w(i_j, p), j = 1, \ldots, k, \forall p\}$.

Let $F(n_1, \ldots, n_k, V)$ be the *partial* flag variety parametrizing flags of quotients of V, with ranks given by the n_i and with universal sequence

(*) $V \otimes \mathcal{O} \twoheadrightarrow Q_{n_k} \twoheadrightarrow \cdots \twoheadrightarrow Q_{n_1}$

There exists a $\phi_{\overline{e}}(t) : \mathcal{U}_{\overline{e}}(t) \to F(n_1, \ldots, n_k, V)$ such that

$$V \otimes \mathcal{O} \twoheadrightarrow K_{i_k}^* Y \cdots \twoheadrightarrow K_{i_1}^*$$

is $\phi_{\overline{e}}(t)^*$ of (*), and $\widetilde{\Omega}_w^{\overline{e}} = \phi_{\overline{e}}^{-1}(t)(D_{w,\overline{e}})$, with $D_{w,\overline{e}}$ defined by the 'same' degeneration condition in $F(n_1, \ldots, n_k, V)$.

LEMMA 2. $D_{w,\overline{e}}$ is irreducible, of codimension a, satisfying $a \geq \ell(w) - \sum e_i$.

LEMMA 3. Let $\overline{e}$ be as above, and assume $\overline{e} \neq 0$. Then

(i) $\sum e_i(e_i - e_{i-1}) \geq 1$;
(ii) We have equality in (i) if and only if $\overline{e} = \overline{e}_{k\ell} = (0, \ldots, 0, \underset{k}{1}, \ldots, \underset{\ell}{1}, 0, \ldots, 0)$ for some $1 \leq k \leq \ell \leq n - 1$;
(iii) Let $\overline{e} = \overline{e}_{k\ell}$ as above, and $w \in S_n$ any permutation, $D_{w,\overline{e}}$ as in Lemma 2. Then $a = \ell(w) - \sum e_i = \ell(w) - (\ell - k + 1)$ if and only if $w(k) > max\{w(k+1), \ldots, w(\ell+1)\}$.

Remark: (ii) and (iii) will be used for the proof of Quantum Monk.

(4): Proof of Theorem 1 (ii). This is done by induction on $\overline{d}$: for $\overline{d} = (0, \ldots, 0)$, $\mathcal{H}Q_{\overline{d}} = H_{\overline{d}} = F$, OK. Assume then $\overline{d} \neq 0$.

Let $c = \sum \ell(w_i)$. It is enough to show that $h_{\bar{e}}(\mathcal{U}_{\bar{e}}) \cap (\cap \Omega_{w_i}(t_i))$ has codimension $> c$ in $\mathcal{H}Q_{\bar{d}}$ for every $\bar{e}$. Now $h_{\bar{e}}$ is birational onto its image, hence it suffices to prove that the codimension of $\cap h_{\bar{e}}^{-1}(\Omega_{w_i}(t_i))$ in $\mathcal{U}_{\bar{e}}$ is greater than

$$c - (\dim \mathcal{H}Q_{\bar{d}} - \dim \mathcal{U}_e) = c + 1 - \sum e_i - \sum e_i(e_i - e_{i-1})$$

By Lemma 1,

$$\cap h_{\bar{e}}^{-1}(\overline{\Omega}_{w_i}(t_i)) = \cap(\pi^{-1}(\mathbb{P}^1 \times \Omega_{w_i}(t_i) \cup \widetilde{\Omega}_{w_i}^{\bar{e}}(t_i)))$$

$\widetilde{\Omega}_{w_i}^{\bar{e}}(t_i)$ is supported on $\mathcal{U}_{\bar{e}}(t_i) = \pi^{-1}(\{t_i\} \times \mathcal{H}Q_{\bar{d}-\bar{e}})$ and $t_1, \ldots, t_N$ are *distinct*, hence there are only two types of nonempty intersections:

(*) $$\cap_{i=1}^N \pi^{-1}(\mathbb{P}^1 \times \overline{\Omega}_{w_i}(t_i))$$

and

(**) $$\cap_{i=1}^{N-1} \pi^{-1}(\mathbb{P}^1 \times \overline{\Omega}_{w_i}(t_i)) \cap \widetilde{\Omega}_{w_N}^{\bar{e}}(t_N)$$

The estimate for (*) is immediate by induction (as π is flat). As for (**), write $W = \cap_{i=1}^{N-1} \pi^{-1}(\{t_N\} \times \overline{\Omega}_{w_i}(t_i))$, so (**)$= W \cap \widetilde{\Omega}_{w_N}^{\bar{e}}(t_N)$. The codimension of W in $\mathcal{U}_{\bar{e}}(t_N)$ is $\sum_{i=1}^{N-1} \ell(w_i) = c - \ell(w_N)$. Now use Kleiman's theorem to deduce that the codimension of (**) in $\mathcal{U}_{\bar{e}}(t_N)$ is $c - \ell(w_N) + a$, and then

$$\mathrm{codim}_{\mathcal{U}_{\bar{e}}}(**) = 1 + c - \ell(w_N) + a \geq 1 + c - \sum e_i > 1 + c - \sum e_i - \sum e_i(e_i - e_{i-1})$$

as needed (the first inequality by Lemma 2, the second by Lemma 3(i)). $\square$

(5): $\langle \Omega_{s_i}, \Omega_w, \Omega_{w'} \rangle_{\bar{d}} =?$ (with $1 + \ell(w) + \ell(w') = \binom{n}{2} + 2\sum d_i$).
This is $\#(\overline{\Omega}_{s_i}(u) \cap \overline{\Omega}_w(v) \cap \overline{\Omega}_{w'}(t))$ for $u, v, t \in \mathbb{P}^1$ distinct; that is,

$$\int_{\mathcal{H}Q_{\bar{d}}} [\overline{\Omega}_{s_i}(u)] \cup [\overline{\Omega}_w(v)] \cup [\overline{\Omega}_{w'}(t)]$$

Suppose that when $u = v$, $Z := \overline{\Omega}_{s_i}(u) \cap \overline{\Omega}_w(u) \cap \overline{\Omega}_w(t)$ is still top-codimensional. Then $\langle \Omega_{s_i}, \Omega_w, \Omega_{w'} \rangle_{\bar{d}} =$ the length of Z. However, Z is supported in the boundary! (this follows from the moving lemma). In fact, one can be much more precise:

PROPOSITION.

(i) Z is either empty, or has pure codimension $\binom{n}{2} + 2\sum d_i$ in $\mathcal{H}Q_{\bar{d}}$;

(ii) Z is contained in $\cup_{\bar{e}_{k\ell}} h_{\bar{e}_{k\ell}}(\mathcal{U}_{\bar{e}_{k\ell}}(u))$, with $\bar{e}_{k\ell}$ as in Lemma 3(ii), and $k \leq i \leq \ell$;

(iii) If $Z \cap h_{\bar{e}_{k\ell}}(\mathcal{U}_{\bar{e}_{k\ell}}(u)) \neq \emptyset$, then w satisfies the condition of Lemma 3(iii): $w(k) > max\{w(k+1), \ldots, w(\ell+1)\}$.

PROOF. The same argument as in the Moving lemma gives that (*) and (**) are empty, as now we have top codimension. One additional case:

$$\tilde{\Omega}^{\bar{e}}_{s_i}(u) \cap \tilde{\Omega}^{\bar{e}}_{w}(u) \cap \pi^{-1}(\{u\} \times \overline{\Omega}_{w'}(t)) \subset \mathcal{U}_{\bar{e}}(u)$$

Since $\ell(s_i) = 1$, we can gain at most one dimension; the strict inequality may become an equality.

(ii), (iii) follow from Lemma 3(ii), (iii). □

(6): Proof of Quantum Monk. Denote by $\alpha_{i,k}$ the product $s_{k-i+1} \cdots s_k$.

$$\langle \Omega_{s_j}, \Omega_{\alpha_{i,k}}, \Omega_{w'} \rangle_{\bar{d}} = \text{length}(Z)$$

The only index $\bar{e}_{k\ell}$ for which $\alpha_{i,k}$ satisfies condition (iii) above is $\bar{e}_k = \bar{e}_{kk}$. Hence, by (ii) above, Z is empty when $j \neq k$.

Assume $j = k$. Then $Z = h_{\bar{e}_k}(W) \subset h_{\bar{e}_k}(\mathcal{U}_{\bar{e}_k}(u))$, with $W = \tilde{\Omega}^{\bar{e}_k}_{s_k}(u) \cap \Omega^{\bar{e}}_{\alpha_{i,k}}(u) \cap \pi^{-1}(\{u\} \times \overline{\Omega}_{w'}(t))$. Recall that $\pi : \mathcal{U}_{\bar{e}_k}(u) \rightarrow \{u\} \times \mathcal{HQ}_{\bar{d}-\bar{e}_k}$ is an open immersion. One checks easily that $\tilde{\Omega}^{\bar{e}_k}_{s_k}(u) = \mathcal{U}_{\bar{e}_k}(u)$ and $\tilde{\Omega}^{\bar{e}}_{\alpha_{i,k}}(u) = \overline{\Omega}_{\alpha_{i-1,k-1}}(u) \cap \mathcal{U}_{\bar{e}_k}(u)$; so $W = \emptyset$ unless $\bar{d} - \bar{e}_k = 0$ and w' is the 'dual' permutation $w_0 \cdot \alpha_{i-1,k-1}$. In this case $\mathcal{U}_{\bar{e}_k}(u) = \{u\} \times F$ and $W = \{pt\}$; moreover $h_{\bar{e}_k}$ is an isomophism onto its image. Hence $Z = \{pt\}$, length $Z = 1$, as needed. □

---

9. Enumerative geometry for hyperelliptic curves—T. Graber
December 5, 1996

Question. How many hyperelliptic curves of genus g, degree d in $\mathbb{P}^2$ pass through $(3d + 1)$ general points?

Strategy. The data of a map from a hyperelliptic curve C to $\mathbb{P}^2$ is almost the same as the data of a map $\mathbb{P}^1 \rightarrow H(2, \mathbb{P}^2)$, the Hilbert scheme of pairs of points in $\mathbb{P}^2$. Lifting such a map:

$$
\begin{array}{ccccc}
C & \longrightarrow & \mathcal{X} & \longrightarrow & \mathbb{P}^2 \\
\downarrow{\scriptstyle 2:1} & & \downarrow{\scriptstyle 2:1} & & \\
\mathbb{P}^1 & \longrightarrow & H(2, \mathbb{P}^2) & &
\end{array}
$$

one gets a honest hyperelliptic curve in $\mathbb{P}^2$ if the bottom map does not land in the diagonal.

Here is the plan:

I. Calculate the genus 0 Gromov-Witten invariants of $H = H(2, \mathbb{P}^2)$;
   —associativity;
   —geometry of H, 'virtual' considerations (H is not convex).
II. Relate the GW-invariants to enumerative geometry;
   —use natural PGL(3) action to control moduli spaces;
   —understand virtual contributions.

Geometry of H. Points of H correspond to either pairs of points in $\mathbb{P}^2$, or points with tangent directions: so there will always be exactly one line containing a given one. This gives a map

$$\pi : H \longrightarrow \mathbb{P}^{2\vee}$$

and $\pi^{-1}([L]) = \mathrm{Sym}^2 L \cong \mathbb{P}^2$. In fact, $H \cong \mathbb{P}(\mathrm{Sym}^2 T\mathbb{P}^2)$. This description allows us to calculate most standard invariants of H. In particular, the Chow ring $A^*(H)$ is generated by divisors:

$T_1 = \pi^*(\mathcal{O}(1))$
$T_2 = \{\text{set of subschemes incident to a fixed line}\}$.

These span the nef cone. A dual basis is given by

$B_1 = \{\text{subschemes supported at a fixed point}\}$;
$B_2 = \text{line in a fiber of } \pi$;

so the effective curves are of the form $(a, b) = aB_1 + bB_2$ with $a, b \geq 0$.

Given a hyperelliptic plane curve C as above, we can recover the degree and genus of C from the homology class (a, b) of the associated rational curve in H. If C has degree d and genus g, then intersecting with T_2 gives $d = b$. To recover the genus of C, note that the branch points of the hyperelliptic involution correspond exactly to the intersections of the rational curve with Δ, the divisor in H parametrizing non-reduced subschemes. It is easy to show that $\Delta = 2(T_2 - T_1)$. We conclude that $2g + 2 = (a, b) \cdot \Delta = 2b - 2a$ so $g = b - a - 1$.

Looking at $H \to \mathbb{P}^2$, the diagonal Δ is realized as a conic bundle inside H: $\Delta = \mathbb{P}(T\mathbb{P}^2) \hookrightarrow \mathbb{P}(\mathrm{Sym}^2 T\mathbb{P}^2)$ by Veronese. The Chow ring $A^*(\Delta)$ is generated by T_1 and $\frac{1}{2}T_2$. Curves in Δ are of the form (a, b) with b even.

Next, $c_1(TH) = 3T_2$. Hence, the expected dimension of $\overline{M}_{0,0}(H, (a, b))$ is $3b + 4 - 3 = 3b + 1$; in particular, it does not depend on the genus of (a, b).

Review of GW invariants:

$$I_\beta(\gamma_1 \cdots \gamma_n) = \int \rho_1^*(\gamma_1) \cup \cdots \cup \rho_n^*(\gamma_n)$$

with usual notations, where the $\int$ is taken on a fundamental class which equals $[\overline{M}_{0,n}(X, \beta)]$ if X is convex. If X is not convex (which is the case at hand), the $\int$ must be taken over a $V \in A_*(\overline{M}_{0,n}(X, \beta)$ of the expected dimension. Heuristically,

$$I_\beta(\cdots) = \#\{\rho_1^{-1}(\Gamma_1) \cap \cdots \rho_n^{-1}(\Gamma_n) \cap V\}$$

with evident notations.

First reconstruction theorem (from [K-M]): if $A^*(X)$ is generated by divisors, then all genus 0 GW-invariants of X can be determined from the 2-point numbers $I_\beta(\gamma_1 \gamma_2)$.

So we look at $I_{(a,b)}(\gamma_1 \gamma_2)$. The class γ_1 imposes at most 3 conditions on curves; two classes impose at most 6 conditions. However, if $b \geq 2$ then the expected dimension is ≥ 7, so the only 2-point numbers come from curve classes $(a, 0)$ or $(a, 1)$.

Start with $(a, 0)$-curves. These lie in Δ, since $(a, 0) \cdot \Delta = -2a < 0$. Curves $(1, 0)$ are fibers of the map $\Delta \to \mathbb{P}^2$ given by support. In other words, all representatives of this class are of the form originally described. So the moduli space $\overline{M}_{0,0}(H, (1, 0))$ is isomorphic to $\mathbb{P}^2$ with universal curve Δ. Curves of type $(a, 0)$ are a-sheeted covers of such fibers. The expected dimension for $(a, 0)$ curves is $3 \cdot 0 + 1 = 1$; so we consider $I_{(a,0)}(\gamma)$ for $\gamma \in A^2(H)$. $A^2(H)$ is 3-dimensional, we'll just look at 2 elements here.

Candidates for γ:

$T_4 = \{$all subschemes incident to a fixed point$\}$;

$T_3 = \{$subschemes contained in a fixed line$\}$.

$I_{(a,0)}(T_4) = I_{(1,0)}(T_4) = 0$ becaues T_4 really imposes two conditions. That is, if we look in $\overline{M}_{0,0}(H, (1, 0)) = \mathbb{P}^2$, the virtual class is an element V of A_1. The locus corresponding to curves meeting a representative of T_4 is just a single point, though, so they don't intersect. Similarly, an $(a, 0)$ curve will meet a representative cycle for T_4 if and only if the $(1, 0)$ curve that it covers meets it, so again the codimension is too high. (There is a second class in $A^2(H)$ which gives a zero GW-invariant for the same reason.)

For T_3 we actually have to worry about the virtual class. It is easy to see that

$I_{(1,0)}(T_3) = \deg(V)$, the degree of the virtual class.

We can actually compute this virtual number (although it will also follow from associativity): consider again $\Delta \subset H$; Δ is made of flags, and hence it is homogeneous. $\overline{M}_{0,n}(\Delta, (a, 0))$ maps isomorphically to $\overline{M}_{0,n}(H, (a, 0))$ since curves of class $(a, 0)$ in H are automatically contained in Δ. We want to find $V \in A_1(\overline{M}_{0,n}(H, (a, 0)))$.

$$\mathcal{U} \xrightarrow{\quad f \quad} \Delta \longrightarrow H$$
$$\downarrow$$
$$\overline{M}_{0,n}(H, (a, 0))$$

By the fancy definition, $V = c_{top}(R^1 \pi_* f^*(N))$. Here N denotes the normal bundle of Δ in H. In the $(1, 0)$ case we have a fine moduli space, so we can actually compute this Chern class using Grothendieck-Riemann-Roch. (The answer is 3.) For $(a, 0)$ with $a \geq 2$ we would have to bring in stacks, and we can do it otherwise anyway.

Next, consider $I_{(a,1)}(\gamma_1 \gamma_2)$. For $a > 1$, $(a, 1) \cdot \Delta < 0$, so have at least a component in Δ. However, the second index is not even, so these curves are all reducible. Curves of type $(a, 1)$ map to lines in $\mathbb{P}^2$; they will consist of a $(0, 1)$ curve, meeting Δ in two points, with a total $(a, 0)$ attached: we glue an $(a_1, 0)$ at the first point and an $(a_2, 0)$ at the second, with $a_1 + a_2 = a$. Because we have such an explicit description of the moduli space, we can identify the space

$$\{\rho_1^{-1}(\Gamma_1) \cap \rho_2^{-1}(\Gamma_2)\}$$

which occurs in the interpretation of the GW-invariants. (Actually we will just identify the image of this locus in the 0-pointed space.)

EXAMPLE. $I_{(a,1)}(\text{pt. class}, T_3)$. The point class should be thought of as a pair of points, $\{p, q\}$ in $\mathbb{P}^2$, and T_3 corresponds to a choice of line l in $\mathbb{P}^2$. Since $\{p, q\}$ is not contained in Δ, the $(0, 1)$ component must hit this point. This determines both the fiber of π in which this line lives, and fixes one point through which it must pass. Another point is determined by the fact that one of the $(a_i, 0)$ components must meet the T_3. As these can be attached only at points of intersection of the $(0, 1)$ curve with Δ, it follows that the curve must contain a double point supported at the intersection of l and $\overline{pq}$. The choice of a_1 and a_2 as well as the choice of particular a_i-sheeted covers has no effect on the incidence relation we are concerned with, except for the condition that a_1 must be non-zero. (a_1 is the degree of the curve glued at the special intersection point.)

The moduli space of solutions splits up into connected components determined by the partition of a between the two intersection points. Once this is decided, all that remains is to choose at each point an a_i sheeted cover of $\mathbb{P}^1$ and a particular point of that cover to glue to the $(0, 1)$ curve. We denote the space of such data by $M(a_i)$. ($M(a)$ is naturally isomorphic to a fiber of the evaluation map from $\overline{M}_{0,1}(\mathbb{P}^1, a)$ to $\mathbb{P}^1$.)

The moduli space of solutions is

$$\amalg_{a_1+a_2=a;a_1>0;a_2\geq 0} M(a_1) \times M(a_2) \quad ;$$

if either a_1 or a_2 is > 1, the corresponding component has positive dimension, and again we need to understand the virtual class in $A_0(M(a_1) \times M(a_2))$. Because the virtual class is constructed from the deformation theory of the stable map, and because the deformations of our $(a, 1)$ curve naturally split into deformations of the $(a_1, 0)$ and $(a_2, 0)$ curves, it follows that this virtual class splits as $V_{a_1} \cdot V_{a_2}$. In conclusion,

$$I_{a,1}(\text{pt}, T_3) = \sum_{a_1+a_2=a, a_1>0, a_2\geq 0} V_{a_1} \cdot V_{a_2}$$

Similar phenomena happen in many cases. In the end, the only unknowns are V_{a_i} and $I_{(a,0)}(T_3)$.

Now, remarkably as usual, the associativity equations are enough to compute

$$\begin{cases} I_{(a,0)}(T_3) = 3/a^2 \\ V_a = \begin{cases} 1 & \text{if } a = 0, 1 \\ 0 & \text{otherwise} \end{cases} \end{cases}$$

From this, all numbers can be computed.

Enumerative significance. The number $I_{(a,b)}(T_4^{3b+1})$ should count hyperelliptic curves through $(3d + 1)$ points. Unfortunately, this is not literally true. Since H is not convex, there will be unwanted contributions to this

GW-invariant. We have unusually good control over these contributions for H however, because of the action of $PGL(3)$. Observe that H is almost homogeneous: the $PGL(3)$ action has only two orbits. So TH is generated by global sections outside Δ. For $f : \mathbb{P}^1 \to H$ with image not contained in the diagonal, $f^*(TH)$ is generically generated by global sections, hence it *is* generated by global sections. Therefore, $H^1(f^*TH) = 0$: H is almost convex. In a neighborhood of such $[f]$, $\overline{M}_{0,n}(\cdots)$ is of the expected dimension. The same holds for reducible curves, as long as no component lies in Δ.

For $f : \mathbb{P}^1 \to \Delta \subset H$ representing a class (a, b), observe that Δ *is* homogeneous; the dimension at such $[f]$ is computed to be $2a + b$. This is $> 3b + 1$ if and only if $a > b$, that is if and only if $(a, b) \cdot \Delta < 0$.

One could hope that if $b > a$, then $\overline{M}_{0,0}(H, (a, b))$ has the expected dimension. However this is not the case, as one can have a curve consisting of two components C_1, C_2, and an f mapping C_2 to Δ and with $f_*[C_1] = (a_1, b_1)$, $f_*[C_2] = (a_2, b_2)$: $b = b_1 + b_2 > a = a_1 + a_2$ doesn't prevent $a_2 > b_2$, in which case $[f]$ moves too much.

However, most of the extraneous components do not contribute. This is because curves in Δ do not hit enough points.

EXAMPLE. $C = C_1 \cup C_2$. Need to hit $3(b_1 + b_2) + 1$ T_4's. Say C_1 moves correctly; then C_1 can only hit $3b_1 + 1$ T_4's. This leaves $3b_2$ for C_2. Now a rational curve in Δ gives rise to a nonreduced subvariety of $\mathbb{P}^2$ supported on a rational curve. The degree of the subscheme is b_2, so this rational curve has degree $b_2/2$. This curve can then hit only $\frac{3}{2}b_2 - 1$ points, $< 3b_2$.

One exception: $b_2 = 0$, $(a, b) = (a_1, b) + (a_2, 0)$. We get solutions here. This is fairly clear. Because the expected dimension doesn't depend on a, and since the dimension of the locus of irreducible curves is equal to the expected dimension, we should see finitely many irreducible curves in class (a_1, b) satisfying the desired incidence conditions for any a_1. Such a curve will meet Δ in $2(b - a_1)$ points. At each of these points you can glue on a $(c_i, 0)$ curve in such a way that $a_1 + \sum c_i = a$. Any such partition of a gives a component of the moduli space which looks like

$$\amalg M(c_1) \times M(c_2) \times \cdots \times M(c_n).$$

The deformation theory for these curves is identical to the deformation theory on the similar moduli spaces we saw earlier. So again the virtual class splits up across the factors, giving a contribution $\sum V_{c_1} V_{c_2} \cdots V_{c_n}$. In fact, all $c_i = 0, 1$ (else $V_{c_i} = 0$). Because of this, we can identify all solutions to the enumerative problem which the GW-invariant is actually solving. Namely, we get the irreducible curves of type (a, b) that we want, but we also get the irreducible solutions of type (a_1, b) for all $a_1 < b$ decorated with $(1, 0)$ curves at exactly $(a - a_1)$ of the points of intersection of the irreducible curve with Δ.

So: define

$$E_{(a,b)}(T_4^{3d+1}) = \#\{\text{irreducible curves of type } (a, b) \text{ meeting the cycles}$$
$$\text{and transverse to } \Delta\}$$

that is the number of honest hyperelliptic curves of degree $d = b$ and genus $b - a - 1$ through $(3d + 1)$ points in $\mathbb{P}^2$. The result is that

$$I_{(a,b)}(T_4^{3d+1}) = \sum_{i=0}^{a} \binom{2b - 2a + 2i}{i} E_{(a-i,b)}$$

and this relation can be inverted to find the $E_{(a,b)}$ in terms of the $I_{(a,b)}$. For example, for genus-2 curves of degree d through $(3d + 1)$ points, we find 27 curves for $d = 4$, 36855 for $d = 5$, 58444767 for $d = 6$, and so on.

For genus 0 and 1, extra care has to be taken to account for 'extra' g_2^1's, as maps $\mathbb{P}^1 \to H$ parametrize a choice of hyperelliptic curve in $\mathbb{P}^2$ *and* a choice of hyperelliptic involution.

---

10. Quantum differential equations and equivariant quantum cohomology, I—R. Pandharipande
December 3, 1996

Example. $\mathbb{C}^*$-equivariant GW-invariants of $\mathbb{P}^2$. Let $\mathbb{C}^*$ act on $\mathbb{P}^2$ by

$$t \mapsto \begin{pmatrix} t^a & & \\ & t^b & \\ & & t^c \end{pmatrix}$$

and let $e_1 = a + b + c$, $e_2 = ab + ac + bc$, $e_3 = abc$. Finite-dimensional approximation of $E\mathbb{C}^* \to B\mathbb{C}^*$: $E\mathbb{C}_n^* = \mathbb{P}(\mathcal{O}(-a) \oplus \mathcal{O}(-b) \oplus \mathcal{O}(-c)) \to B\mathbb{C}_n^* = \mathbb{P}^n$. A module basis of the (ordinary) equivariant cohomology of $\mathbb{P}^2$ over $H_{\mathbb{C}^*}^* = \mathbb{C}[t]$ is

$$\begin{cases} 1 \\ T_1 = \xi = c_1(\mathcal{O}_{\mathbb{P}}(1)) \\ T_2 = \xi^2 \end{cases}$$

The equivariant pairing matrix is

$$(g_{ef}) = \begin{pmatrix} 0 & 0 & 1 \\ 0 & 1 & e_1 t \\ 1 & e_1 t & (e_1^2 - e_2)t^2 \end{pmatrix}$$

So we can compute (g^{ef}). Proceed as usual, setting up a 'quantum equivariant potential'

$$\Gamma = \sum_{d > 0} \sum_{k \geq 0} I_d(T_2^{3d-1+k}) e^{dy_1} \frac{y_2^{3d-1+k}}{(3d - 1 + k)!}$$

By dimension reasoning, $I_d(T_2^{3d-1+k}) = N_{d,k} t^k$ with $N_{d,k}$ numbers. Associativity as in the usual story: read the coefficient of T_0 in $(T_1 * T_1) * T_2 = T_1 * (T_1 * T_2)$ and obtain

$$\Gamma_{222} = \Gamma_{112}^2 - \Gamma_{111}\Gamma_{122} + 2e_1 t\Gamma_{122} - (e_1^2 + e_2)t^2\Gamma_{112} + (e_1 e_2 - e_3)t^3\Gamma_{111}$$

Then find equations for the $N_{d,k}$'s: for $d > 0$, $k \geq 0$, $(d,k) \neq (1,0)$,

$$
N_{d,k} = \sum_{d_1,d_2 \geq 1; k_1, k_2 \geq 0; d_1 + d_2 = d; k_1 + k_2 = k} N_{d_1,k_1} N_{d_2,k_2} d_1 d_2.
$$

$$
\cdot \left[d_1 d_2 \binom{3d - 4 + k}{3d_1 - 2 + k_1} - d_1^2 \binom{3d - 4 + k}{3d_1 - 1 + k_1} \right]
$$

$$
+ 2e_1 d N_{d,k-1} - (e_1^2 + e_2) d^2 N_{d,k-2} + (e_1 e_2 - e_3) d^3 N_{d,k-3}
$$

The old equation is just a subrecursion here ($k = 0$; set $N_{d,k} = 0$ if $k < 0$). This equation determines all the numbers recursively, from $N_{1,0} = 1$. For example, $N_{1,1} = 2e_1$.

Note: the $N_{d,k}$ are symmetric functions of degree k in a, b, c.

Dubrovin formalism. Let $\mathcal{S}$ be a trivial bundle over $M = \mathbb{R}^n$, with sections $\hat{s}_1, \ldots, \hat{s}_m$ which trivialize it. We consider a connection $\nabla : H^0(\mathcal{S}) \to H^0(\mathcal{S} \otimes \Omega_M^1)$, that is a map on C^∞ sections, satisfying

$$
\nabla(f \cdot s) = f \nabla s + s \otimes df
$$

for f a function. Vector fields $\frac{\partial}{\partial x_i}$ determine covariant derivatives $\nabla_i = \nabla_{\frac{\partial}{\partial x_i}}$ by contracting ∇s by $\frac{\partial}{\partial x_i}$.

In local coordinates, we may write

$$
\nabla(\hat{s}_j) = \Gamma_{ij}^k \hat{s}_k \otimes dx_i
$$

(omitting obvious $\sum$), and for $g^j \hat{s}_j \in H^0(\mathcal{S})$

$$
\nabla_i (g^j \hat{s}_j) = \frac{\partial g^j}{\partial x_i} \hat{s}_j + \Gamma_{ij}^k g^j \hat{s}_k
$$

We can think of a connection as a way of lifting vector fields from M to $\mathcal{S}$ (that is, a *distribution* in the sense of Frobenius): for each point $s \in \mathcal{S}$, we have n independent vectors in $T_s \mathcal{S}$:

$$
\frac{\partial}{\partial x_p} \mapsto \frac{\partial}{\partial x_p} - \Gamma_{pj}^k s_j \frac{\partial}{\partial s_k}
$$

(in coordinates $x_1, \ldots, x_n, s_1, \ldots, s_m$ on $\mathcal{S}$).

Now we want sections s for which $\nabla s = 0$. We get Frobenius integrability conditions: if $\mathcal{L}_1$, $\mathcal{L}_2$ are vector fields in the distribution, then $[\mathcal{L}_1, \mathcal{L}_2]$ must also be in it. This is where the curvature comes up: denoting by L_p the lift of $\frac{\partial}{\partial x_p}$ is $[L_p, L_q]$ the lift of a vector field? Computation:

$$
[L_p, L_q] = \left\{ -\frac{\partial}{\partial x_p} \Gamma_{qj}^k + \frac{\partial}{\partial x_q} \Gamma_{pj}^k + \Gamma_{pj}^\ell \Gamma_{q\ell}^k - \Gamma_{qj}^\ell \Gamma_{p\ell}^k \right\} s_j \frac{\partial}{\partial s_k}
$$

No x-term, so the part in $\{\ \}$ must be 0 for integrability. This term is denoted by R^k_{jpq}, the coefficients of the curvature form in local coordinates.

Dubrovin. Apply this to $M = V = H^*(X)$ (say $\mathbb{C}$ coefficients), with basis $T_0, \ldots, T_m$, and $S = T_V$: the trivial bundle with fiber V. The coordinates on the manifold V will be denoted y_i, and ∂_i will be coordinates on the fiber. There is a metric on T_V: $(g_{ij}) = \langle \partial_i, \partial_j \rangle$. Define a connection by setting

$$\nabla_i \partial_j = \Phi_{ije} g^{(ek)} \partial_k$$

that is by prescribing Christoffel symbols $\Gamma^k_{ij} = \Phi_{ije} g^{ek}$. Here Φ is the GW potential, that is

$$\sum_{n \geq 3} \frac{1}{n!} \sum_\beta I_\beta(\gamma^{\otimes n})$$

Now compute the curvature (R^k_{jpq}) and impose its vanishing (that is, the integrability condition): $\frac{\partial}{\partial y_p} \Gamma^k_{qj} = \frac{\partial}{\partial y_q} \Gamma^k_{pj}$ comes for free from the definition, and the rest says

$$\Phi_{pje} g^{e\ell} \Phi_{q\ell f} = \Phi_{qje} g^{e\ell} \Phi_{p\ell f}$$

(times the invertible matrix (g^{fh})). This are just the WDVV equations!

So at least formally there are parallel sections. Givental (in one part of [Givental]) writes down such sections. His goal in doing this: operators that kill these sections will give relations in QH^*.

Givental. $\nabla_\hbar = \hbar d - \sum(p_\alpha *) dt_\alpha \wedge$

Here $\hbar$ is just a parameter; $p_1, \ldots, p_n$ form a basis of $H = H^*X$, and for its (trivial) tangent bundle T_H (same notation). $\nabla_\hbar$ acts $H^0(T_H) \to H^0(T_H \otimes \Omega^1)$.

We will obtain sections $\gamma = \sum t_i p_i$, with coordinates $(t) = (t_1, \ldots, t_n)$. Consider vector fields $F \in H^0(T_H)$, in coordinates:

$$F(t) = \begin{pmatrix} F_1 \\ \vdots \\ F_n \end{pmatrix} = \sum F^j p_j$$

Covariant derivatives:

$$\nabla_{\hbar, i}(F(t)) = \sum_k \left(\hbar \frac{\partial F^k}{\partial t_i} - \Phi_{ije} g^{ek} F^j \right) p_k$$

(so $\nabla_{\hbar, i}(p_j) = -p_i * p_j$, the quantum product). If $\hbar = 1$, $\nabla_\hbar$ is a honest connection; else, strictly speaking, $\nabla_\hbar$ is only a connection "up to scalar"; but the same Frobenius equations hold.

Consider now $\mathbb{P}^1 = \mathbb{PC}^2$; $\mathbb{C}^*$ acts by $\begin{pmatrix} t^0 & \\ & t^1 \end{pmatrix}$. Equivariant cohomology:

$H^*_{\mathbb{C}^*} \mathbb{P}^1 = \mathbb{C}[p, \hbar]/(p^2 - p\hbar)$. In terms of the basis $\{1, p\}$, $(g_{ef}) = \begin{pmatrix} 0 & 1 \\ 1 & \hbar \end{pmatrix}$. Given polynomials $f(p,\hbar)$, $g(p,\hbar)$, we have an equivariant pairing $\langle f,g \rangle \in H^*_{\mathbb{C}^*} = \mathbb{C}[\hbar]$. In fact, one checks that

$$\langle f,g \rangle = \frac{1}{2\pi i} \int \frac{fg\,dp}{p(p-\hbar)}$$

Over $\mathbb{C}(\hbar)$ we can take the basis $\frac{p}{\hbar}$, $\frac{\hbar-p}{\hbar}$, for which the intersection form diagonalizes to $\begin{pmatrix} \frac{1}{\hbar} & 0 \\ 0 & -\frac{1}{\hbar} \end{pmatrix}$.

Back to our manifold X, $V = H^*X$, ∇ on T_V. We want sections s with $\nabla s = 0$. Look then at $X \times \mathbb{P}^1$, $\mathbb{C}^*$ acting trivially on X and as above on $\mathbb{P}^1$; and use equivariant GW invariants on $X \times \mathbb{P}^1$. The equivariant cohomology is

$$H^*_{\mathbb{C}^*}(X \times \mathbb{P}^1) = H^*X \otimes_{\mathbb{C}} \mathbb{C}[p,\hbar]/(p(p-\hbar))$$

Look for a $\mathbb{C}[\hbar]$-module basis: we have $\{p_i\}$ for H^*X; and over $\mathbb{C}(\hbar)$, we have $p_i \otimes \frac{p}{\hbar}$, $p_i \otimes \frac{\hbar-p}{\hbar}$. Write

$$\varphi = \text{``}(\tau,t)\text{''} = \sum t_i \left(p_i \otimes \frac{p}{\hbar} \right) + \tau_i \left(p_i \otimes \frac{\hbar-p}{\hbar} \right)$$

for elements of $H^*_{\mathbb{C}^*}(X \times \mathbb{P}^1) \otimes_{\mathbb{C}[\hbar]} \mathbb{C}(\hbar)$. Then the equivariant intersection pairing is nice:

$$\langle \varphi, \varphi' \rangle = \langle (\tau,t),(\tau',t') \rangle = \frac{\langle t,t' \rangle - \langle \tau,\tau' \rangle}{\hbar}$$

where $\langle t,t' \rangle = \langle \sum t_i p_i, \sum t'_i p_i \rangle$ and $\langle \tau,\tau' \rangle = \langle \sum \tau_i p_i, \sum \tau'_i p_i \rangle$ are both calculated on X alone.

In fact, for more equivariant classes, the equivariant push-forward is given by

$$\int \varphi \cup \varphi' \cup \varphi'' = \frac{\left(\int_X t \cup t' \cup t'' \right) - \left(\int_X \tau \cup \tau' \cup \tau'' \right)}{\hbar}$$

Here τ's and t's do not mix because $p(\hbar-p)$ in the numerator kills the poles, so the residue is 0.

Next, consider the following function G of $(t,\tau,\hbar,q,q_0)$:

$$G = \sum_{n \geq 0} \sum_{\beta} \sum_{d} \frac{1}{n!} q^\beta q_0^d I^{\text{equiv.}}_{n,(\beta,d)}(\varphi^{\otimes n})$$

Here (β,d) is a curve class of $X \times \mathbb{P}^1$ in the evident fashion, and φ is written in terms of $\hbar$ etc., as above. Note that the $q^\beta q_0^d$ terms 'pull apart' the curve classes; that's what Givental always does. Write

$$G = G^{(0)} + q_0 G^{(1)} + q_0^2 G^{(2)} + \dots$$

CLAIM. $G^{(0)} = \dfrac{\Phi(t,q) - \Phi(\tau,q)}{\hbar}$, with $\Phi = $ ordinary GW-potential on X.

This comes down to the preceding equations for the equivariant push–forward. Another basis, e.g. without $1/\hbar$, would not simplify so much.

Next, set $\theta_{\alpha\beta} = \dfrac{\partial^2 G^{(1)}}{\partial\tau_\alpha \partial\tau_\beta}$.

CLAIM. Fix β. Set $F = \sum_\alpha \theta_{\alpha\beta} g^{\alpha j} p_j$. Then

$$-\hbar \frac{\partial}{\partial\tau_\gamma} F = p_\gamma(\tau) * F$$

$$\hbar \frac{\partial}{\partial t_\gamma} F = p_\gamma(t) * F$$

(Warning: typo in [Givental]: t and τ are swapped.)

Note: the entries of F involve p_α's, and $p_\gamma * p_\alpha = \sum$ power series(var.)p_f. "$p_\gamma(\tau)$" means: set the variable to τ in the power series. That is, say $F = \sum F^i p_i$; now $p_\gamma * \sum F^i p_i = \sum F^i p_\gamma * p_i$; $p_\gamma * p_i = \sum \Phi_{\gamma ie} g^{ef} p_f$. So we can define $p_\gamma(\tau) * F$ to be $\sum_{i,e,f} F^i \Phi_{\gamma ie}(\tau) g^{ef} p_f$, and similarly for $p_\gamma(t) * F$.

The F given by the last Claim are the 'parallel sections', but they involve τ's and t's.

To prove the claim, one essentially uses the fact that G satisfies the equivariant WDVV equation for $X \times \mathbb{P}^1$.

---

11. Quantum differential equations and equivariant quantum cohomology, II—R. Pandharipande
December 10, 1996

Review of the previous lecture: X is a variety; $V = H^*(X, \mathbb{C})$; we have a natural identification $V \cong T_V$; $\{p_1, \ldots, p_N\}$ is a basis of V, and $\partial_1, \ldots, \partial_N$ the corresponding basis of T_V; we let t_i be coordinates on V. We have defined a connection 'up to scalar':

$$\nabla_\hbar = \hbar d - \sum_\alpha (p_\alpha *) dt_\alpha \wedge$$

acting in the following way: for a vector field $F(t) = \begin{pmatrix} F_1 \\ \vdots \\ F_N \end{pmatrix} = \sum F^j p_j = \sum F^j \partial_j$, we have the covariant derivative

$$\nabla_{\hbar,i} F = \sum_k \left(\hbar \frac{\partial F^k}{\partial t_i} - \Phi_{ije} g^{ek} F^j \right) \partial_k$$

where Φ denotes the GW-potential.

We have seen that Φ satisfies the WDVV equations $\iff$ the corresponding formal connection is flat. We are looking for solutions of $\nabla_{\hbar,i} F = 0$ $\forall i$; that is, $\forall i, k$

$$\hbar \frac{\partial}{\partial t_i} F^k = \Phi_{ije} g^{ek} F^j$$

Givental's approach: to get solutions, consider $X \times \mathbb{P}^1$ with $\mathbb{C}^*$ acting on the $\mathbb{P}^1$ factor by $(t^0 : t^1)$. Fact:

$$H^*_{\mathbb{C}^*} \mathbb{P}^1 = \mathbb{C}[p, \hbar]/(p^2 - p\hbar)$$

We found a better basis here, over $\mathbb{C}(\hbar)$: $p/\hbar$, $(\hbar - p)/\hbar$.

REMARK. By *localization*, the equivariant cohomology of $\mathbb{P}^1$ is essentially concentrated at $0, \infty$; this basis is the natural one from this point of view.

We then have a basis for $H^*_{\mathbb{C}^*}(X \times \mathbb{P}^1) = H^* X \otimes \mathbb{C}[p, \hbar]/(p^2 - p\hbar)$: $p_i \otimes p/\hbar$, $p_i \otimes (\hbar - p)/\hbar$. So we write $\varphi \in H^*_{\mathbb{C}^*}(X \times \mathbb{P}^1)$ as

$$\varphi = \sum t_i \left(p_i \otimes \frac{p}{\hbar} \right) + \sum \tau_i \left(p_i \otimes \frac{\hbar - p}{\hbar} \right)$$

Next, consider the *equivariant* GW-potential $G(t, \tau, \hbar, q, q_0)$; we expand it in terms of powers of the $\mathbb{P}^1$-curve class:

$$G = G^{(0)} + q_0 G^{(1)} + q_0^2 G^{(2)} + \cdots$$

We have seen that

(I) $G^{(0)} = \dfrac{\Phi(t, q) - \Phi(\tau, q)}{\hbar}$, the usual GW-potential;

(II) letting $\theta_{ab} := \dfrac{\partial^2 G^{(1)}}{\partial \tau_a \partial t_b}$ and $F := \sum_a \theta_{ab} g^{aj} \partial_j$, then we have the two equations

$$\begin{cases} -\hbar \dfrac{\partial}{\partial \tau_\gamma} F = \partial_\gamma(\tau) * F \\[2mm] \hbar \dfrac{\partial}{\partial t_\gamma} F = \partial_\gamma(t) * F \end{cases}$$

(end of the review)

Localization. Let Y be a manifold with a $\mathbb{C}^*$ action, and denote by $Y^{\mathbb{C}^*}$ the set of fixed points. We have $\pi_* : H^*_{\mathbb{C}^*}(Y) \to H^*_{\mathbb{C}^*}$ as usual, and also $H^*_{\mathbb{C}^*}(Y^{\mathbb{C}^*}) \cong H^*(Y^{\mathbb{C}^*}) \otimes H^*_{\mathbb{C}^*}$ since the action is trivial on $Y^{\mathbb{C}^*}$. The equivariant inclusion $Y^{\mathbb{C}^*} \hookrightarrow Y$ gives a diagram

$$\omega \in H^* \mathbb{C}^*(Y) \longrightarrow \omega \in H^*_{\mathbb{C}^*}(Y^{\mathbb{C}^*}) = H^*(Y^{\mathbb{C}^*}) \otimes H^*_{\mathbb{C}^*}$$

$$\pi_* \omega \in H^*_{\mathbb{C}^*}$$

The diagonal arrow can be filled in: it works $\omega \mapsto \int_{Y^{\mathbb{C}^*}} \frac{\omega}{c_{top}^{eq} N}$, where $N = $ normal bundle of $Y^{\mathbb{C}^*}$ in Y.

That is, $\pi_* \omega$ can be evaluated on fixed points. In this sense, equivariant cohomology is 'concentrated at the fixed points'.

Equivariant GW-invariants.

$$I_{n,d+1} \left\langle p_a \left(\frac{\hbar - p}{\hbar} \right) \cdot p_b \left(\frac{p}{\hbar} \right) \cdot \varphi^{\otimes n} \right\rangle$$

is the equivariant push-forward of ω from $Y = \overline{M}_{0,n+2}(X \times \mathbb{P}^1, d+1)$, where "$(d+1)$" stands for the sum of a class d from the X factor and of the class of a point from $\mathbb{P}^1$, and (with evident notations)

$$\omega = e_1^* \left(p_a \left(\frac{\hbar - p}{\hbar} \right) \right) \cup e_2^* \left(p_b \left(\frac{p}{\hbar} \right) \right) \cup e^*(\varphi)^{n \text{ times}}$$

The idea is to compute the fixed point locus $Y^{\mathbb{C}^*}$, then use localization to compute the push-forward.

What is $Y^{\mathbb{C}^*}$? We want the class to be 1 on the $\mathbb{P}^1$ factor, so a map $\mathbb{P}^1 \to X \times \mathbb{P}^1$ must travel up exactly once; and must go straight up for a fixed point (else a surface is swept). Next,

$$X \times \mathbb{P}^l$$

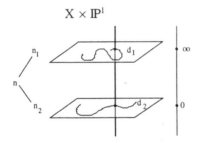

$p_a((\hbar - p)/\hbar)$ must sit on the ∞-plane, and $p_b(p/\hbar)$ must sit on 0; the other $n = n_1 + n_2$ points can be split arbitrarily among the two planes. This describes $Y^{\mathbb{C}^*}$.

Count the conditions: two nodes (along the vertical component), fixed at 0 and ∞; total of four conditions, so expect codim$_Y$ $Y^{\mathbb{C}^*}$ to be 4. As for the normal bundle, the tangent representation is t at ∞, corresponding to $-\hbar$; and t^{-1} at 0, corresponding to $\hbar$; smoothing the nodes, we get $-\hbar + c(\infty)$ at 0, and $\hbar + c(0)$ at ∞. The localization formula gives then the push-forward as

$$\int_{Y^{\mathbb{C}^*}} \frac{\omega}{-\hbar^2(-\hbar + c(\infty))(\hbar + c(0))}$$

This is all modulo a slight lie: $Y^{\mathbb{C}^*}$ is not codimension 4 everywhere. All n points may be concentrated on one of the planes, leaving no nodes at the other plane: the two possibilities give two codimension 3 components, with normal bundles whose classes are resp.

$$-\hbar^2(\hbar + c(0)) \quad , \quad -\hbar^2(-\hbar + c(\infty))$$

Putting all together, $I_{n,d+1}\langle\cdots\rangle$ is the sum of three terms (one from cod. 4, two from cod. 3), involving only 'ordinary' GW-invariants of X. The first term is

$$\sum \frac{1}{-\hbar^2}\frac{n!}{n_1!n_2!}I_{n_1+2,d_1}\left\langle p_a \cdot \tau^{\otimes n_1}\frac{p_\epsilon}{-\hbar + c(\infty)}\right\rangle g^{\epsilon\epsilon'} I_{n_2+2,d_2}\left\langle p_b \cdot t^{\otimes n_2}\frac{p_{\epsilon'}}{-\hbar + c(0)}\right\rangle$$

where the $\sum$ is over $d_1 + d_2 = d$; $n_1 + n_2 = n$; $(d_1, n_1) \neq (0,0)$, $(d_2, n_2) \neq (0,0)$; the $g^{\epsilon\epsilon'}$ factor makes sure the vertical component goes straight up: the point at 0 equals the point at ∞. The second and third term are similar:

$$\frac{1}{-\hbar^2}g_{ae}g^{\epsilon\epsilon'}I_{n+2,d}\left\langle p_b \cdot t^{\otimes n}\frac{p_{\epsilon'}}{\hbar + c(0)}\right\rangle$$

$$+ \frac{1}{-\hbar^2}I_{n+2,d}\left\langle p_a \cdot \tau^{\otimes n}\frac{p_\epsilon}{-\hbar + c(\infty)}\right\rangle g^{\epsilon\epsilon'}g_{\epsilon'b}$$

To simplify notations, define a new function:

$$\psi_{ef}(t,\hbar) = g_{ef} + \sum_{n\geq 0,(n,d)\neq(0,0)}\sum_d \frac{1}{n!}I_{n+2,d}\left\langle \frac{p_t}{\hbar + c}\cdot t^{\otimes n}\cdot p_a\right\rangle$$

Then in short localization gives

$$-\hbar^2\theta_{ab} = \sum_{\epsilon,\epsilon'}\psi_{ae}(\tau, -\hbar)g^{\epsilon\epsilon'}\psi_{be'}(t,\hbar)$$

(note: typos in [Givental]).

Now recall that θ must satisfy equations (II) (from the review):

$$-\hbar\frac{\partial}{\partial\tau_\gamma}\theta_{ab}g^{ak}\partial_k = p_\gamma(\tau)*\theta_{ab}g^{ak}\partial_k$$

Substituting the expression obtained above into this:

$$-\hbar\frac{\partial}{\partial\tau_\gamma}\psi_{ae}(\tau, -\hbar)g^{\epsilon\epsilon'}\psi_{\beta e'}(t,\hbar)g^{ak}\partial_k = p_\gamma(\tau)*\psi_{ae}(\tau, -\hbar)g^{\epsilon\epsilon'}\psi_{be'}(t,\hbar)g^{ak}\partial_k$$

Taking out the invertible part:

$$-\hbar\frac{\partial}{\partial\tau_\gamma}\psi_{ae}(\tau, -\hbar)g^{\alpha k}\partial_k = p_\gamma(\tau)*\psi_{ae}(\tau, -\hbar)g^{\alpha k}\partial_k$$

or, after $\hbar \leftrightarrow -\hbar$:

$$\hbar \frac{\partial}{\partial \tau_\gamma} \psi_{a\epsilon}(\tau, \hbar) g^{\alpha k} \partial_k = p_\gamma(\tau) * \psi_{a\epsilon}(\tau, \hbar) g^{\alpha k} \partial_k$$

From this, one can obtain n independent solutions.

A different approach is obtained via *Gravitational descendents*, going back to Witten. We have defined ordinary GW-invariants by

$$\int_{[\overline{M}_{g,n}(X,\beta)]} e_1^*(\gamma_1) \cup \cdots \cup e_n^*(\gamma_n)$$

with notations as usual (in particular, with due care if X is non convex, $g > 0$, etc.). We can more generally define invariants

$$\int_{[\overline{M}_{g,n}(X,\beta)]} e_1^*(\gamma_1) \cup \cdots \cup e_n^*(\gamma_n) \cup c_1^{\alpha_1} \cup \cdots \cup c_n^{\alpha_n}$$

where c_i =first Chern class of i-th cotangent line. Notation:

$$I_{n,\beta} \langle (\gamma_1, \alpha_1) \cdots (\gamma_n, \alpha_n) \rangle = \int_{[\overline{M}_{g,n}(X,\beta)]} \prod e_i^*(\gamma_i) \cup c_i^{\alpha_i}$$

The ordinary GW-invariants satisfy properties:

(I) *Fundamental class:* $I_{n+1,\beta} \langle \gamma_1 \cdots \gamma_n \cdot 1 \rangle = 0$ if $n \geq 3$ or $\beta \neq 0$, $n \geq 0$;
(II) *Divisor:* for $\gamma \in H^2(X)$, $I_{n+1,\beta} \langle \gamma_1 \cdots \gamma_n \cdot \gamma \rangle = \int_\beta \gamma \cdot I_{n,\beta} \langle \gamma_1 \cdots \gamma_n \rangle$.

Now the fancier invariants satisfy upgraded properties:

(I) *Fundamental class:*

$$I_{n+1,\beta} \langle (\gamma_1, \alpha_1) \cdots (\gamma_n, \alpha_n)(1, 0) \rangle = \sum_{i=1}^{n} I_{n,\beta} \langle (\gamma_1, \alpha_1) \cdots (\gamma_i, \alpha_i - 1) \cdots (\gamma_n, \alpha_n) \rangle$$

(of course, setting to 0 terms involving negative α-components);
(II) *Divisor:* for $\gamma \in H^2(X)$,

$$I_{n+1,\beta} \langle (\gamma_1, \alpha_1) \cdots (\gamma_n, \alpha_n)(\gamma, 0) \rangle = \int_\beta \gamma \cdot I_{n,\beta} \langle (\gamma_1, \alpha_1) \cdots (\gamma_n, \alpha_n) \rangle$$

$$+ \sum_{i=1}^{n} I_{n,\beta} \langle (\gamma_1, \alpha_1) \cdots (\gamma_i \cdot \gamma, \alpha_i - 1) \cdots (\gamma_n, \alpha_n) \rangle$$

The underlying reason why these hold: consider the map forgetting the $(n+1)$-st point:

$$\nu : \overline{M}_{0,n+1}(X,\beta) \to \overline{M}_{0,n}(X,\beta)$$

At the first point, we have c_1, $\nu^* c_1$: one can show that

$$c_1 = \nu^* c_1 + [D_{1,n+1}]$$

where $D_{1,n+1}$ is the divisor obtained by splitting (on ≥ 2 components) the points as $(1, n+1)|(2,\ldots,n)$, and the class as $0|\beta$.

For $g=0$, the fancier invariants also satisfy suitable WDVV equations. As in the ordinary GW-case, the splitting axiom and the linear equivalence in $\overline{M}_{0,4}$ yield recursive relations for the invariants.

The claim is that the fundamental class and divisor properties, plus the upgraded WDVV, imply that the function $\psi_{ef}(t,\hbar)$ defined above is a solution of the main differential equation. This gives a second, and non-equivariant proof, of the main theorem.

In [Dubrovin], all this is seen from an axiomatic point of view. There a reconstruction theorem for $g=0$ is stated:

PROPOSITION. *The tree-level gravitational descendents can be uniquely reconstructed from the tree-level system of Gromov-Witten invariants.*

SKETCH OF PROOF. Seek an inductive scheme

$$I_{n,\beta} \langle (\gamma_1, \alpha_1) \cdots (\gamma_n, \alpha_n) \rangle$$

$$= \sum \text{(either lower curve class, or fewer cotangent line classes)}$$

For this: the gravitational descendents are obtained by integrating

$$e_1^*(\gamma_1) c_1^{\alpha_1} \cdots e_n^*(\gamma_n) c_n^{\alpha_n}$$

over $\overline{M}_{0,n}(X,\beta)$. Instead, integrate

$$e_1^*(\gamma_1) c_1^{\alpha_1} \cdots e_n^*(\gamma_n) c_n^{\alpha_n} \cup e_{n+1}^*(H) \cup e_{n+2}^*(H) \cup e_{n+3}^*(1)$$

(with H =hyperplane class) over $\overline{M}_{0,n+3}(X,\beta)$. If we are not yet done, then some of the α_i is $\neq 0$, say α_n. Apply then the basic linear equivalence $D(12|34) \sim D(14|23)$ with "1"= $e_n^*(\gamma_n) c_n^{\alpha_n}$, "2"= $e_{n+1}^*(H)$, "3"= $e_{n+2}^*(H)$, and "4"= $e_{n+3}^*(1)$. This makes the induction click. □

---

12. Quantum double Schubert polynomials—W. Fulton
December 12, 1996

This is joint work with Ciocan-Fontanine.

Reminder on quantum Schubert polynomials (after Fomin–Gel'fand–Postnikov): for $w \in S_n$, we can define a polynomial $\mathfrak{S}_w^q(x)$, as follows. Quantum elementary symmetric polynomials are defined by $E_i(x_1,\ldots,x_p) := \sum \text{products}$

of vertices and edges which cover exactly i vertices once in

$$
\begin{array}{ccccc}
x_1 & x_2 & x_3 & \cdots & x_p \\
\bullet \quad q_1 & \bullet \quad q_2 & \bullet \quad q_3 & \cdots \quad q_{p-1} & \bullet
\end{array}
$$

($\deg x_i = 1, \deg q_i = 2$). Clearly $E_i(x_1, \ldots, x_p)$ specialize to the usual elementary symmetric polynomials $e_i(x_1, \ldots, x_p)$ as the $q_j \mapsto 0$.

DEFINITION. Write the usual Schubert polynomial corresponding to $w \in S_n$ as

$$\mathfrak{S}_w(x) = \sum a_J e_J(x)$$

with $a_J \in \mathbb{Z}$, and denoting $e_J(x) = e_{j_1}(x_1) e_{j_2}(x_1, x_2) \cdots e_{j_{n-1}}(x_1, \ldots, x_{n-1})$, with $0 \le j_i \le i$. Set analogously $E_J(x) = E_{j_1}(x_1) \cdots E_{j_{n-1}}(x_1, \ldots, x_{n-1})$; then define

$$\mathfrak{S}_w^q(x) = \sum a_J E_J(x)$$

The theorem ('Quantum Giambelli', see the previous lecture on quantum Schubert polynomials) is that these $\mathfrak{S}_w^q(x)$ represent the Schubert varieties in the quantum cohomology of flag manifolds.

Now for the classical Schubert polynomials there is a concrete description: write $w = w_0 s_{i_1} \cdots s_{i_\ell}$ in the shortest possible way, where $w_0 = n(n-1) \cdots 1$ is the 'longest' permutation, s_i is the transposition $(i, i+1)$, and $\ell = \ell(w_0) - \ell(w)$. Then

$$\mathfrak{S}_w(x) = \partial_{i_\ell}^x \circ \cdots \circ \partial_{i_1}^x (\mathfrak{S}_{w_0}(x))$$

where $\mathfrak{S}_{w_0}(x) = x_1^{n-1} x_2^{n-2} \cdots x_1$ and $\partial_i^x(P) = \dfrac{P - s_i^x(P)}{x_i - x_{i-1}}$ (see also previous lectures). The goal here is to find a similar formula for $\mathfrak{S}_w^q(x)$.

The idea comes from the following notion. There exist 'double Schubert polynomials' $\mathfrak{S}_w(x, y)$, obtained as

$$\mathfrak{S}_w(x, y) = \partial_{i_\ell}^x \circ \cdots \circ \partial_{i_1}^x (\mathfrak{S}_{w_0}(x, y))$$

with $\mathfrak{S}_{w_0}(x, y) = \prod_{i+j \le n}(x_i + y_j)$ (note: in the usual definition, one finds $-$ instead of $+$). It would seem as if the y_j are useless; on the contrary:

FACT. $\mathfrak{S}_w(x, y) = \mathfrak{S}_{w^{-1}}(y, x)$.

Equivalently, write $w = s_{j_\ell} \cdots s_{j_1} w_0$ again in the shortest possible way; then

$$\mathfrak{S}_w(x, y) = \partial_{j_\ell}^y \circ \cdots \circ \partial_{j_1}^y (\mathfrak{S}_{w_0}(x, y))$$

DEFINITION. With w as above, define *quantum double Schubert polynomials* by

$$\mathfrak{S}_w(x, y, q) = \partial_{j_\ell}^y \circ \cdots \circ \partial_{j_1}^y (\mathfrak{S}_{w_0}(x, y, q))$$

where

$$\mathfrak{S}_{w_0}(x, y, q) = \prod_{p=1}^{n-1} E_p(x_1 + y_{n-p}, \ldots, x_p + y_{n-p})$$

Note that $\mathfrak{S}_w(x, y, 0) = \mathfrak{S}_w(x, y)$ by definition. The interesting fact is that

THEOREM. $\mathfrak{S}_w(x, 0, q) = \mathfrak{S}_w^q(x)$.

PROOF. Claim: $\mathfrak{S}_w(x, y, q) = \sum a_J(y) E_J(x)$ with $a_J(y) \in \mathbb{Z}[y]$.
Granted this, $\mathfrak{S}_w(x, 0, q) = \sum a_J(0) E_J(x)$, and the $a_J(0)$ are the right co-efficients to give $\mathfrak{S}_w^q(x)$ since $\mathfrak{S}_w(x) = \mathfrak{S}_w(x, 0, 0) = \sum a_J(0) e_j(x)$.
So we only need to prove the Claim.

PROOF OF THE CLAIM. (1) True for $w = w_0$: use that

$$E_p(x_1 + y_{n-p}, \ldots, x_p + y_{n-p}) = \sum_{i=0}^{p} E_i(x_1 \ldots x_p)(y_{n-p})^{p-i}$$

which is clear from the definition.
(2) The ∂_i^y's only work on the coefficients $a_J(y)$. $\square$

Geometric origin. On the hyperquot scheme$\times \mathbb{P}^1$ we have

$$E_1 \subset \cdots \subset E_{n-1} \subset V \to Q_{n-1} \to \cdots \to Q_1$$

where the E_i's are subbundles of V, the Q_i's are bundles, but the maps on the right are not necessarily surjective. We want the locus Ω'_w where heuristically

$$\text{"rk}(E_p \to Q_q) \leq \#\{i \leq q, w(i) \leq p\}\text{"}$$

If each $V \to Q_i$ is surjective, then $[\Omega'_w] = \mathfrak{S}_w(x, y)$, where x_i, y_i denote (up to sign) the class of successive quotients of the Q_j's, E_j's respectively. When the maps are not surjective, higher Chern classes of the Q_i's must be used.
For Ω'_{w_0}, we need the locus where $E_1 \to Q_{n-1}$ vanishes; this is

$$c_{top}(E_1^\vee \otimes Q_{n-1}) = \sum_{i=0}^{n-1} c_i(Q_{n-1}) y_1^{n-1-i} = E_{n-1}(x_1 + y_1, \ldots, x_{n-1} + y_1)$$

On this locus, look further where $E_2/E_1 \to Q_{n-2}$ vanishes. A computation gives that this is $E_{n-2}(x_1 + y_2, \ldots, x_{n-2} + y_2)$, and so on.

Remark: here $x_1 + \cdots + x_i = c_1(Q_i)$, and $q_i = c_2(Q_{i+1} - Q_i)$; all other Chern classes are determined by c_1's and c_2's. This must be a special property of the universal bundles on the hyperquot scheme.

Part III
Related Material

1. Mirror symmetry and string-theoretic
Hodge numbers—V. Batyrev
September 12, 1996

Problems arise in trying to define and compute $QH^*(X)$ when X is singular: for example, for the 'small' QH^* one does not expect the variables $q_1, \ldots, q_r$ to be defined as generators of $H^2(X)$; one expects more parameters, corresponding to a larger (conjectural) 'string-theoretic' $H_{st}(X)$. This lecture deals with the problem of defining mirror symmetry for singular varieties, and more specifically with suitable Hodge numbers displaying the symmetry.

§1. Mirror symmetry. First a reminder of a basic example. Let A be the abelian group $(\mathbb{Z}/(n+2)\mathbb{Z})^{\oplus n+2}$, of order $(n+2)^{n+2}$. Consider the nondegenerate bilinear form $< \cdot, \cdot >: A \times A \to \mathbb{Z}/(n+2)\mathbb{Z}$ given by

$$\langle (\alpha_0, \ldots, \alpha_{n+1}), (\beta_0, \ldots, \beta_{n+1}) \rangle = \sum \alpha_i \beta_i \mod (n+2)$$

B denotes the diagonal $\mathbb{Z}/(n+2)\mathbb{Z} \hookrightarrow A$; note that B is isotropic: $\langle B, B \rangle = 0$. Define $G = B^{\perp}/B$, another abelian group, of order $(n+2)^n$; and restrict the pairing to $G \times G \to \mathbb{Z}/(n+2)\mathbb{Z}$.

Now move to algebraic geometry. Consider the Fermat hypersurface $\phi_n \subset \mathbb{P}^{n+1}$ defined by $\sum z_i^{n+2} = 0$; G acts naturally on ϕ_n, by $(\alpha, z_i) \mapsto e^{2\pi i \alpha} z_i$.

STATEMENT. *For all $H \subset G$, and denoting by $H^{\perp}$ the complement of H with respect to the pairing in G, the physicists claim that*

$$(\phi_n/H) \leftrightarrow (\phi_n/H^{\perp})$$

are mirror-symmetric.

EXAMPLE. $n = 3$, $H = 0$: then $G = (\mathbb{Z}/5\mathbb{Z})^3$, and the claim is that ϕ_5 and ϕ_5/G are mirror-symmetric.

Note: in this case ϕ_5 is smooth, but ϕ_5/G is singular.

Another statement: if V, V' are smooth n-dimensional Calabi-Yau, and mirror symmetric, then $h^{n-p,q}(V) = h^{p,q}(V')$ for all p, q.

How to state this if V or V' are singular? First natural conjecture: the equality of Hodge numbers should hold for some minimal desingularization. We expect that ϕ_n/H, $\phi_n/H^\perp$ have "nice" desingularizations displaying mirror symmetry of Hodge numbers. In fact, this has been proved for $n = 3$.

How to control the 'minimality' of a desingularization?

DEFINITION. X Gorenstein variety, K_X canonical divisor. A resolution $\pi : Y \to X$ is *crepant* if $\pi^* K_X = K_Y$.

(In general, $K_Y = \pi^* K_X + \sum a_i E_i$, with E_i the components of the exceptional divisors. The a_i's are called 'discrepancies'; so $a_i = 0 \leftrightarrow$ there are *no* discrepancies $\leftrightarrow$ the resolution is crepant. This terminology is due to Miles Reid.)

Note: if $Z \subset Y$ (both nonsingular) and $Y' = Bl_Z Y$, with exceptional divisor E, then $K_{Y'} = \pi^* K_Y + (\text{codim } Z - 1)E$. Unless Y' is isomorphic to Y, codim $Z - 1 > 0$; so 'crepancy' guarantees the minimality of the resolutions in the sense that blowing-up inessential loci will not give crepant desingularizations.

§2. **E-polynomials.** Assume that X is quasi-projective. Then the cohomology with compact support $H_c^*(X)$ has a mixed Hodge structure, and we may define

$$e^{p,q}(X) = \sum_k (-1)^k h^{p,q}(H_c^*(X))$$

DEFINITION. $E(X; u, v) = \sum_{p.q} e^{p,q}(X) u^p v^q$.

Properties:
  (1) If $X = \amalg X_i$, then $E(X; u, v) = \sum E(X_i; u, v)$
  (2) $Y \to X$ locally trivial in the Zariski topology, with fiber $F \implies$
      $E(Y; u, v) = E(F; u, v) \cdot E(X; u, v)$.

(Note: (2) does fail unless one has Zariski-local triviality.)

Now assume that $\pi : Y \to X$ is a resolution of singularities of X, with Y smooth and compact. Then $e^{p,q}(Y) = (-1)^{p+q} h^{p,q}(Y)$ by the purity of the Hodge structure of Y; so knowing $E(Y; u, v)$ amounts to knowing the Hodge numbers of Y. On the other hand, say that we have a stratification $X = \amalg X_i$ with the analytic singularity of X constant along X_i, and $\pi^{-1}(X_i) \to X_i$ Zariski-locally trivial; then one ought to be able to compute $E(Y; u, v)$.

QUESTION. Assume that $\pi : Y \to X$ is a crepant resolution. For $x \in X$, what is $E(\pi^{-1}(x); u, v)$?

This question can be 'localized' for quotient singularities as follows: for $G \subset SL(n, \mathbb{C})$, $X = \mathbb{C}^n/G$, let $\bar{0} \in X$ be the image of $0 \in \mathbb{C}^n$. If $\pi : Y \to X$ is a minimal desingularization, what is $E(\pi^{-1}(\bar{0}); u, v)$?

An answer to this could be pasted into an answer to the global question if X has quotient singularities and $Y \to X$ is crepant. Note that crepant resolutions do not necessarily exist:

EXAMPLE. $n = 4$, $G \cong \mathbb{Z}/2\mathbb{Z}$ generated by $(-1, -1, -1, -1)$. Then $\mathbb{C}^n/G$ does not have crepant resolutions. (Crepant resolutions do exist for $n < 4$.)

However, if two crepant resolutions exist then their E-polynomial will have to be the same (see below).

§3 Orbifold Euler number. (Dixon, Vafa, Witten, ...) Say a group G acts on a variety X; then

$$e(X, G) = \frac{1}{|G|} \sum_{(g,h):gh=hg} e(X^g \cap X^h)$$

is the 'orbifold' or *string-theoretic* Euler number, denoted $e_{\mathrm{st}}(X/G)$.

Note: the conventional Euler number of the quotient would instead be $e(X/G) = \frac{1}{|G|} \sum_g e(X^g)$. The string-theoretic Euler number should encode more information.

EXAMPLE. $e_{\mathrm{st}}(\mathbb{C}^n/G) = \#$ of conjugacy classes, while $e(\mathbb{C}^n/G) = 1$.

Now for V, V' smooth n-dimensional mirror-symmetric Calabi-Yau, one has $e(V) = (-1)^n e(V')$. The physicists propose that for quotient varieties this should hold for d_{st}. Supporting that this is the right notion, we have

THEOREM. $e_{\mathrm{st}}(\phi_n/H) = (-1)^n e_{\mathrm{st}}(\phi_n/H^\perp)$

Now back to the local question from §2.

EXPECTATION. $E(\pi^{-1}(\bar{0}); u, v) = \sum \#(\textit{conj. classes with weight } j)(uv)^j$

Here the *weight* of a conjugacy class is defined as follows: for $g \in G$, write g in terms of roots of 1: $g = \mathrm{diag}(e^{2\pi i\alpha_1}, \ldots, e^{2\pi i\alpha_n})$ with $0 \le \alpha_j < 1$. We are assuming $G \subset SL$, so $e^{2\pi i \sum \alpha_j} = 1$, and hence $\sum \alpha_j$ is a nonnegative integer. We let the weight of g be this integer.

It is not hard to check that the 'expectation' is correct for 2-dimensional quotient singularities. The exceptional divisor over a singularity is then a chain of say m rational curves, and one checks that the E-polynomial of such a chain is indeed $1 + (m - 1)uv$ as prescribed by the above formula.

Also, the 'expectation' makes sense regardless of whether a crepant resolution should exist or not. For the singularity $\mathbb{C}^4/(-1, -1, -1, -1)$ we get $1 + (uv)^2$; and more generally, for $\mathbb{C}^{2n}/\mathbb{Z}^2$ one finds $1 + (uv)^n$.

Finally, for 'toric resolutions' one can formulate a precise results:

THEOREM. *For toric (complete intersection, Calabi-Yau) mirror pairs V, V':*

$$u^n E_{\mathrm{st}}(V; u^{-1}, v) = (-1)^n E_{\mathrm{st}}(V'; u, v)$$

where E_{st} is computed from the 'expected' E-polynomial of fibers of a resolution, as given above.

If we denote by $h_{\mathrm{st}}^{p,q}(V)$ the coefficient of $u^p v^q$ in E_{st}, the Theorem formulates then this case of mirror symmetry as the equality of 'string-theoretic' Hodge numbers

$$h_{\mathrm{st}}^{p,q}(V) = h_{\mathrm{st}}^{p,q}(V')$$

§4. **Number theory (p-adic fields).** (Only very rough ideas here)
The 'expectation' is a true formula!

To prove this, one may either use an approach via infinite dimensional geometry (suggested by M. Kontsevich), or p-adic integration theory on $\mathbb{Q}_p$. This is analogue to Lebesgue integration on $\mathbb{R}$ or $\mathbb{C}$, with Haar measure normalized by $\mu(\mathbb{Z}_p) = 1$.

A. Weil: the integral of a volume form along the maximal compact subset $X(\mathbb{Z}_p)$ of a variety X defined over $\mathbb{Q}_p$ will be $\dfrac{\#X(\mathbb{Z}_p)}{|p|^{\dim X}}$.

Now say that V, V' are birational. As integration is insensitive to proper subvarieties, and V, V' are biregularly isomorphic away from proper subvarieties, we will get the same number of points for V, V' over $\mathbb{Z}_p$. Via the Weil conjectures, this in some sense explains why one should expect the same cohomological properties for V and V'. If $Y \to X$ is a crepant resolution, a volume form Ω on X will extend with same zeros and poles on Y independently of the specific crepant resolution (by definition of crepant), and from this one can understand why one should get the same E for different resolutions.

§5. **Extending to more general singularities.** Suppose X has log terminal singularities. Can we define a string-theoretic E-polynomial for X?

Let $\pi : Y \to X$ be *any* resolution, with $D_1, \ldots, D_r$ exceptional divisors. Write $K_Y = \pi^* K_X + \sum_{i \in I} a_i D_i$, with $I = 1, \ldots, r$. For $J \subset I$, define

$$D_J = \cap_{j \in J} D_j \quad , \quad D_J^\circ = D_J - \cap_{i \in J} D_i$$

and $D_\emptyset = Y - \cup_{i \in J} D_j$.

Then $Y = \amalg_{J \subset I} D_J^\circ$. We define then

DEFINITION.

$$e_{\mathrm{st}}(X) = \sum_{J \subset I} e(D_J^\circ) \left(\prod_{j \in J} \frac{1}{a_j + 1} \right)$$

This agrees with the previous e_{st} for quotient singularities.

THEOREM. *This number does not depend on the resolution.*

It is immediate to check that blowing up a given resolution at a point, for example, does not change the number defined above. The present proof of this Theorem however is not so naive, and again uses p-adic integration.

This e_{st} seems the right candidate to use in a more general formulation of mirror symmetry.

---

2. $\overline{M}_{0,n}$—P. Belorousski
September 17, 1996

References for this material are [Knudsen], [Keel], and (for a different construction) [Kapranov].

We denote by $M_{0,n}$ the space parametrizing ordered n-tuples of distinct points on $\mathbb{P}^1$ modulo projective transformations; we work over $\mathbb{C}$.

§1. **Naive compactifications of** $M_{0,n}$. The first observation is that any three points of any ordered n-tuple of distinct points in $\mathbb{P}^1$ (say, the last three) can be placed at 0, 1, ∞ by means of a unique projective transformation. The position of the other $n - 3$ points determines then the n-tuple up to projective transformation, with the only constraint that they should be distinct from each other, and distinct from 0, 1, ∞.

$n = 3$: $M_{0,3} = \cong$point;

$n = 4$: $M_{0,4} \cong \mathbb{P}^1 - \{0, 1, \infty\}$;

more generally, $M_{0,n} \cong (\mathbb{P}^1 - \{0, 1, \infty\})^{n-3}-$ diagonals, or alternatively $M_{0,n} \cong (\mathbb{C} - \{0, 1\})^{n-3}-$ diagonals.

Naive compactifications of these spaces are then $(\mathbb{P}^1)^{n-3}$, or $\mathbb{P}^{n-3}$; neither of these is adequate in the sense that points at the boundary are not 'geometrically meaningful'. For example, the natural action of S_n on $M_{0,n}$ does not extend to an action on these compactifications.

EXAMPLE. Consider the 5-tuple of points

$$(p_1, p_2, p_3, p_4, p_5) = (0, 1, \infty, t, t^2)$$

for $t \neq 0$, then let t approach 0. In the naive compactifications, p_4 and p_5 both approach p_1. However, if we take p_1, p_4, p_5 to be the points fixed at $0, 1, \infty$ by applying $z \mapsto \frac{z(1-t)}{z-t^2}$, the 5-tuple becomes

$$(p_1, p_2, p_3, p_4, p_5) = (0, \tfrac{1}{1+t}, 1 - t, 1, \infty)$$

and as t approaches 0 this time the points p_2, p_3 and p_4 come together. That is, the geometric interpretation of the limiting configuration depends on the choice of three fixed points.

§2. **The moduli problem and Knudsen's construction.**

DEFINITION. A stable n-pointed curve of genus 0 is a connected projective nodal curve C with n distinct smooth marked points p_i, with dim $H^1(C, \mathcal{O}_C) = 0$, and such that each component of C has at least 3 special points.

Here *special* means singular or marked. We will denote an n-pointed curve by $(C, p_1, \ldots, p_n)$.

The stability condition implies that the curve has no nontrivial automorphisms fixing the marked points.

REMARK. For a connected nodal curve,

$$\text{genus} = \sum g_i + (\#\text{nodes}) - (\#\text{components}) + 1$$

where g_i denotes the genus of the i-th irreducible component. Therefore genus$= 0 \implies g_i = 0$ for all i, and $(\#\text{nodes}) - (\#\text{components}) = 1$. In particular, all components of C are necessarily copies of $\mathbb{P}^1$.

In fact, these components form a tree. It is convenient to introduce the *dual graph* of a stable curve, by setting

$$\text{vertices} = \{\text{components of } C\}$$
$$\text{edges} = \{\text{nodes}\}$$
$$\text{tails} = \{\text{marked points}\}$$

In this terminology, a 'vertex' is an *internal* vertex of the graph, while a 'tail' is a boundary vertex, with an edge attaching it to the rest of the graph. For example, here is a schematic representation of a 7-pointed stable curve and of the associated graph:

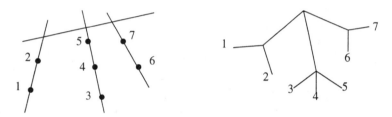

In terms of the graph of C, genus$= 0 \implies$ the graph is a tree; and stability $\implies$ the valence (i.e. the number of edges attached to it) of each internal vertex is at least 3.

DEFINITION. A *family* of n-pointed stable curves of genus 0 over a base scheme S is a flat projective morphism $C \xrightarrow{\pi} S$ with n sections $\sigma_1, \ldots, \sigma_n$: $S \to C$ such that

$$(C_s, \sigma_1(s), \ldots, \sigma_n(s))$$

is a reduced stable n-pointed curve of genus 0 for all $s \in S$. Here C_s denotes the fiber of C over $s \in S$,

Morphisms of families over a given base scheme are defined in the obvious way; the moduli problem is defined by the (contravariant) functor associating to each scheme the set of equivalence classes (up to isomorphism) of all families defined over it.

THEOREM. *There exists a projective smooth algebraic variety* $\overline{M}_{0,n}$, *which is the fine moduli space for n-pointed stable genus-0 curves.*

That is, for all schemes S there is a natural bijection

$$\left\{ \begin{array}{c} \text{Isom. classes of families of } n\text{-pointed} \\ \text{stable curves of genus zero over } S \end{array} \right\} \leftrightarrow \text{Hom}(S, \overline{M}_{0,n})$$

We sketch the proof given by Knudsen.

PROOF SKETCH. Induction on n. For $n = 3$, $\overline{M}_{0,3} \cong$ point, since 3-pointed stable curves are necessarily irreducible. The universal family $\mathcal{C}_3 \to \overline{M}_{0,3}$ will be $\mathbb{P}^1$, with 3 marked points.

Key observation: $\mathcal{C}_n \cong \overline{M}_{0,n+1}$!

For the construction of the universal family, look first at the case $n = 4$. We have the diagram

$$
\begin{array}{ccccc}
\mathcal{C}_4 & \longrightarrow & \mathcal{C}_3 \times_{\overline{M}_{0,3}} \mathcal{C}_3 & \xrightarrow{\cong} & \mathbb{P}^1 \times \mathbb{P}^1 \\
\downarrow & & \downarrow & & \downarrow \\
\overline{M}_{0,4} & = & \overline{M}_{0,4} = \mathcal{C}_3 & \xrightarrow{\cong} & \mathbb{P}^1
\end{array}
$$

There are three obvious sections on $\mathcal{C}_3 \times_{\overline{M}_{0,3}} \mathcal{C}_3$; a fourth one is given by the diagonal:

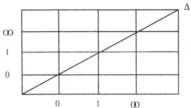

the fibers are 4-pointed curves, but we have to separate the diagonal from the sections at the points $(0,0)$, $(1,1)$, (∞,∞); we blow-up $\mathcal{C}_3 \times_{\overline{M}_{0,3}} \mathcal{C}_3$ at the three intersection points, and this produces $\mathcal{C}_4$.

Note that the effect of the blow-up is to sprout out a new component (an exceptional divisor) on which the "glued" points will separate.

The inductive step for higher n is analogous. Assuming $\mathcal{C}_n \to \overline{M}_{0,n}$ has been constructed, take $\overline{M}_{0,n+1}$ to be $\mathcal{C}_n$, start by considering $\mathcal{C}_n \times_{\overline{M}_{0,n}} \mathcal{C}_n$, and blow up the intersections of the diagonal with the basic sections and with the singular loci of the fibers. The centers of blow-up might not be regularly embedded, so one will need to resolve the singularities introduced by blow-up; a minimal desingularization will yield $\mathcal{C}_{n+1}$. □

The fact that $\overline{M}_{0,n}$ is a fine moduli space yields

COROLLARY 1. S_n acts on $\overline{M}_{0,n}$ by permuting the points.

Also, we may index the points by arbitrary finite sets: if $|A| < \infty$, denote by $\overline{M}_{0,A}$ the corresponding $\overline{M}_{0,|A|}$. From Knudsen's proof one can derive

COROLLARY 2. *Given $B \subset A$, $|A| < \infty$, there is a contraction morphism* $\overline{M}_{0,A} \to \overline{M}_{0,B}$.

EXAMPLE. Say $A = \{1, \ldots, n\}$, $B = \{1, \ldots, n-1\}$; we get a morphism $\overline{M}_{0,n} \to \overline{M}_{0,n-1}$ forgetting the n-th point. If the n-th point is on a component with only 3 special points, just forgetting it would destabilize the component; in practice, the effect of the operation is to contract such a component:

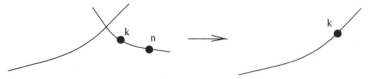

Of course, the algebra behind this is not as simple as the set-theoretic description.

REMARK. Changing the order in which you forget two different points gives two possible compositions

$$\overline{M}_{0,n} \to \overline{M}_{0,n-1} \to \overline{M}_{0,n-2}$$

The resulting maps $\overline{M}_{0,n} \to \overline{M}_{0,n-2}$ must however coincide, since they clearly agree on the open locus $M_{0,n}$.

Taking this observation further, we see that for any 4-tuple of distinct indices i,j,k,ℓ in $\{1,\dots,n\}$ there is a (unique) map

$$\overline{M}_{0,n} \to \overline{M}_{0,\{i,j,k,\ell\}} \cong \mathbb{P}^1$$

forgetting all the other points.

§3. Geometry of $\overline{M}_{0,n}$.

FACT. *The boundary of the compactification* $M_{0,n} \hookrightarrow \overline{M}_{0,n}$ *is a divisor with normal crossings.*

For all partitions $A \amalg B = \{1,\dots,n\}$ with $|A| \geq 2$, $|B| \geq 2$, we have a divisor $D(A|B)$ at the boundary, whose general point corresponds to a curve of the type

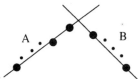

The points at the 'boundary' of $D(A|B)$ correspond to possible degenerations:

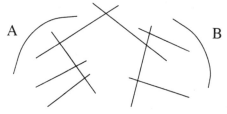

The intersection of any number of these $D(A|B)$'s is either empty or smooth. $\overline{M}_{0,n}$ has a stratification by locally closed loci indexed by the combinatorial type of the degeneration.

More precisely, the *combinatorial type* of a curve is simply its dual graph (with marked tails).

EXAMPLE. In $\overline{M}_{0,9}$ (of dimension 6):

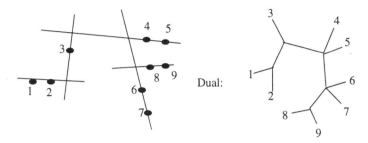

Dual:

All curves with this combinatorial type form a stratum in $\overline{M}_{0,9}$. The closure of this stratum is

$$D(12|3\cdots 9) \cap D(123|4\cdots 9) \cap D(1\cdots 5|6\cdots 9) \cap D(1\cdots 7|89)$$

The boundary cycles (i.e. the closures of the strata) are themselves products of $\overline{M}_{0,k}$'s for $k < n$. For example,

$$D(A|B) \cong \overline{M}_{0,A\cup\{*\}} \times \overline{M}_{0,\{*\}\cup B}$$

where we think of $*$ as the points at which the A-stable curve and the B-stable curve are glued.

In other words, for every k, ℓ there is a morphism

$$\overline{M}_{0,k+1} \times \overline{M}_{0,\ell+1} \hookrightarrow \overline{M}_{0,k+\ell}$$

and the image of this map is one of the boundary divisors in the target.

EXAMPLES. $\overline{M}_{0,4} \cong \mathbb{P}^1$:

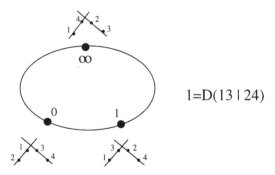

$1=D(13 \mid 24)$

$\overline{M}_{0,5}$: blow-up $\mathbb{P}^1 \times \mathbb{P}^1$ at 3 general points (as seen in Knudsen's proof in the last section). Equivalently, one can blow-up $\mathbb{P}^2$ at 4 general points, since blowing-up $\mathbb{P}^2$ at two points is the same as blowing-up $\mathbb{P}^1 \times \mathbb{P}^1$ at 1 point.

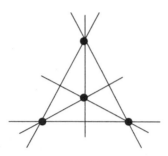

One gets the same answer by considering Procesi's compactification for this configuration. There will be 4 exceptional divisors, plus the proper transforms of the six lines through the points, which will give 6 more (-1)-curves. The blow-up of $\mathbb{P}^2$ at four general points is a Del Pezzo surface of degree five, embedded into $\mathbb{P}^5$ by the system of cubics through the four points. The 10 special curves become lines in this embedding.

These 10 lines are the boundary divisors of $\overline{M}_{0,5}$, which are indexed as $D(A|B)$ by the $\binom{5}{2} = 10$ subsets A, $|A| = 2$, of $\{1, \ldots, 5\}$. The incidence of the $D(A|B)$'s is represented by the following graph:

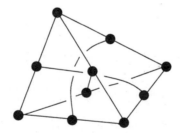

Finally, $\overline{M}_{0,5}$ is the universal curve over $\overline{M}_{0,4} \cong \mathbb{P}^1$:

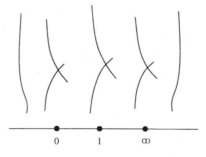

the 3 special fibers give 6 exceptional curves; plus the 4 sections, for a total of 10.

§4. **Chow ring of $\overline{M}_{0,n}$.** We can obtain some geometrically transparent relations in the ring as follows:

(a) $D(A|B) = D(B|A)$;

(b) the choice of distinct i, j, k, ℓ in $\{1, \ldots, n\}$ gives a map $p : \overline{M}_{0,n} \to \overline{M}_{0,4} \cong \mathbb{P}^1$;

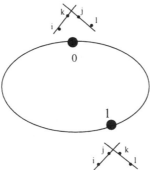

$D(ij|k\ell) = D(ik|j\ell)$ up to rational equivalence in $\mathbb{P}^1$, hence

$$p^* D(ij|k\ell) = p^* D(ik|j\ell)$$

in $A^*\overline{M}_{0,n}$. That is,

$$\sum_{i,j \in A; k, \ell \in B} D(A|B) = \sum_{i,k \in A; j, \ell \in B} D(A|B)$$

(c) $D(A|B) \cdot D(C|D) = 0$ unless the curves corresponding to the general points in the divisors have common degenerations, that is unless one of the four sets A, B, C, D contains one of the others (i.e., the common refinement of the two partitions consists of 3 rather than 4 sets). For example, $A = \{1, 2\}$, $B = \{3, 4, 5\}$, $C = \{1, 3\}$, $D = \{2, 4, 5\}$:

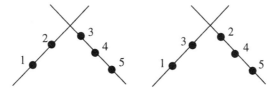

there is no common degeneration to

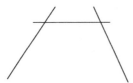

so $D(12|345) \cdot D(13|245) = 0$.

The $D(A|B)$'s generate the Chow ring of $\overline{M}_{0,n}$ multiplicatively. Surprisingly:

THEOREM. *(a), (b), (c) are the only relations in the Chow ring of $\overline{M}_{0,n}$. Therefore,*

$$A^*\overline{M}_{0,n} = \mathbb{Z}[D(A|B), A \amalg B = \{1,\ldots,n\}, |A| \geq 2, |B| \geq 2]/((a),(b),(c))$$

Also, $H^*\overline{M}_{0,n} \cong A^*\overline{M}_{0,n}$.

These results are due to Sean Keel, [Keel].

---

3. Quantum cohomology, old and new—Z. Ran
September 26, 1996

LEMMA. *Consider a 3-dimensional irreducible closed subvariety V of the $\mathbb{P}^N$ parametrizing plane curves of a given degree d. Suppose V parametrizes rational curves: i.e., a general point $C \in V$ corresponds to an irreducible rational curve. Then V has a degenerate member: there exists a reducible or multiple $C_0 \in V$.*

Note that the bound '3' is sharp: the $\mathbb{P}^2$ of lines has no degenerate members.

PROOF. Pick general points a, b in $\mathbb{P}^2$, and consider a one-parameter sub-family B of V consisting of curves through a, b. The total space of B gives

$$X \xrightarrow{\;F\;} \mathbb{P}^2$$
$$\pi \downarrow$$
$$B$$

where F is a morphism and $X \to B$ is a blown-up $\mathbb{P}^1$-bundle. We may assume that F is relatively minimal, i.e., there are no vertical (-1)-curves E such that $F(E) =$pt. The general fiber of π is a $\mathbb{P}^1$; the special fibers are trees of $\mathbb{P}^1$'s. Each of these has at least two ends; these are (-1)-curves, so F is not constant along them.

Claim: there must exist at least one degenerate fiber.

Indeed, otherwise π is a $\mathbb{P}^1$-bundle; there would be sections $F^{-1}(a) = S_a$, $F^{-1}(b) = S_b$. Let then $L = F^*(\text{line})$: as F is generically finite, $L^2 > 0$; $L \cdot S_a = 0$; $L \cdot S_b = 0$. By the Hodge index theorem, $S_a^2 < 0$ and $S_b^2 < 0$. But $S_a - S_b$ is a multiple of the class of the fiber, hence

$$0 = (S_a - S_b)^2 = S_a^2 + S_b^2 < 0 \quad,$$

contradiction. □

Now we could try to get a formula from this situation for $\deg V = \deg F$, by relating it to the number of degenerate fibers containing either one or both of S_a, S_b. Using ideas from quantum cohomology, one should get

$$d - \deg V + \sum_W \deg_2 W \cdot (\deg C_2)^2 = \sum_W \deg_{1,1} W \cdot (\deg C_1) \cdot (\deg C_2)$$

where W ranges over 'boundary components' of V: W parametrizes curves of the form $C_1 \cup C_2$ where

either C_1 is a 2-parameter family, and $\deg_2 W = \#\{C_1 \cup C_2\colon C_1$ contains 2 general points$\}$,

or $\{C_1\}$ and $\{C_2\}$ are each a 1-parameter family, and $\deg_{1,1} W = \#\{C_1 \cup C_2 : C_1$ contains a general point, and C_2 contains a general point$\}$.

So much for the 'new' methods; move now to the 'old' method, degenerating the target.

$V_{d,\delta}$ denotes the Severi variety of nodal plane curves of degree d with δ nodes. $V_{d,\delta}$ is locally closed in $\mathbb{P}^{\binom{d+2}{2}-1}$; its dimension is

$$\binom{d+2}{2} - 1 - \delta = 3d + g - 1 \quad ,$$

where $g = \frac{(d-1)(d-2)}{2} - \delta$ is the genus. We denote by $N_{d,\delta}$ the degree of $V_{d,\delta}$.

Consider the blow-up

$$S = B\ell_{(0,p)}\mathbb{C} \times \mathbb{P}^2 \longrightarrow \mathbb{C} \times \mathbb{P}^2$$

$$\pi \downarrow$$

$$\mathbb{C}$$

π is a flat family, and $\pi^{-1}(t) = \mathbb{P}^2$ for $t \neq 0$; $\pi^{-1}(0) = S_0 = S_1 \cup S_2$, with $S_1 = B\ell_p\mathbb{P}^2$, and $S_2 = \mathbb{P}^2$.

The intersection $E = S_1 \cap S_2$ sits in S_1 as the exceptional curve, and in S_2 as a line.

The blow-up sits in $(\mathbb{C} \times \mathbb{P}^2) \times \mathbb{P}^2$; so there is another map $S \to \mathbb{P}^2$. We have

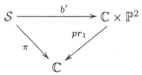

with $b'|_{S_2} =$ identity, and $b'|_{S_1} =$ a $\mathbb{P}^1$-bundle over image $= E \cong \mathbb{P}^1$.

Now consider $V_{d,\delta}$ as a family of curves on the general fiber of π, and take the limit as $t \to 0$. As a cycle,

$$V_{d,\delta} \to \sum m(\pi)V(d, e/\pi/d, \delta_2/\delta_1)$$

where: π is a partition, $\pi = [\ell_1, \ldots, \ell_r]$, $\ell_i = \#$ of blocks of size i; the weight of π is $|\pi| = \sum i\ell_i$, and here $e = |\pi|$; $m(\pi) = \prod i^{\ell_i}$; the $\sum$ is over π, δ_1, δ_2 satisfying

$$\sum(i - 1)\ell_i + \delta_1 + \delta_2 = \delta$$

The different π correspond to the strata of divisors of degree $|\pi|$ in $\mathbb{P}^1$: $\pi \mapsto \sum_{i=1}^{r} \sum_{j=1}^{d_i} iQ_{ij}$.

$V(d, e/\pi/d, \delta_2/\delta_1)$ is the set of all $C_1 \cup C_2$ such that $C_i \subset S_i$ has δ_i nodes, and is smooth near E; $C_1 \cap E = C_2 \cap E = $ divisor of type π in $E \cong \mathbb{P}^1$.

Note: the C_i are not necessarily irreducible—this makes the degrees easier to compute.

Next, the sought degree $N_{d,\delta}$ is the number of curves $C \in V(d, \delta)$ containing N general points $p_1, \ldots, p_N$, where $N = \frac{(d+1)(d+2)}{2} - 1 - \delta$. Specialize $p_1, \ldots, p_N$ to N_1 points on S_1 and N_2 points on S_2, $N_1 + N_2 = N$. To avoid trivial solutions, assume

$N_1 \geq d + 1$ (avoids limit components where $C_1 = d$ rulings)

$N_2 \geq 3$.

One choice: make N_2 as large as possible, i.e., choose $N_1 = d + 1$. In this case, either $e = d - 1$ or else C contains a component not meeting E; call it $C_{1,0}$; this would come from a proper component of C itself, so C must be reducible. This cannot happen if $\delta < d$.

For the rest of the discussion, assume $e = d - 1$. Then $C_1 \in |dH - (d-1)E|$ on S_1, and

$$C_1 = \sum_{i=1}^{\delta_1} R_i + C_{1,0}$$

with R_i rulings, and $C_{1,0} \in |(d - \delta_1)H - (d - \delta_1 - 1)E|$ a smooth rational curve. (One can in fact view this as a degeneration in the plane.)

We need to express the condition $C_1 \cap E = C_2 \cap E$. Let P_π be the set of divisors of type π on $E = \mathbb{P}^1$, and $P^\pi = \prod \mathbb{P}^{\ell_i}$. P^π maps birationally to P_π. The condition $C_1 \cap E = C_2 \cap E$ is equivalent to

$$(C_1 \cap E, C_2 \cap E) \in \Delta_1 \times \cdots \times \Delta_r \subset P^\pi \times P^\pi = (\mathbb{P}^{\ell_1} \times \mathbb{P}^{\ell_1}) \times \cdots \times (\mathbb{P}^{\ell_r} \times \mathbb{P}^{\ell_r})$$

Now

$$(*) \ \Delta_1 \times \cdots \times \Delta_r \sim \left(\sum_{j=0}^{\ell_1} \mathbb{P}^j \times \mathbb{P}^{\ell_1 - j} \right) \times \cdots \times \left(\sum_{j=0}^{\ell_r} \mathbb{P}^j \times \mathbb{P}^{\ell_r - j} \right) = \sum_{\pi' + \pi'' = \pi} P^{\pi'} \times P^{\pi''}$$

where $\pi' = [\ell_1', \ldots]$, $\pi'' = [\ell_1'', \ldots]$, and $\pi' + \pi'' = [\ell_1' + \ell_1'', \ldots]$.

Write $C_m \cap E = \sum \sum iQ_{ij}^m$, $m = 1, 2$. Fix Q_{ij}^1, $j = 1, \ldots, \ell_i'$, and Q_{ij}^2, $j = \ell_i' + 1, \ldots, \ell_i' + \ell_i'' = \ell_i$. Define

$V_{(d,e),\delta,\backslash \pi',\pi''} = \{$nodal curves C_1 with δ nodes, and

$\quad\quad\quad\quad C_1 \cap$ line $= D_{\pi'}^{\text{fixed}} + D_{\pi''}$, $D_{\pi''}$ of general type π'', $|\pi'| + |\pi''| = e\}$

There is an analogous locus $V_{e,\delta,/\pi',\pi''}$. Also, we set $N_{d,\delta,/\pi',\pi''} = \deg V_{d,\delta,/\pi',\pi''}$, etc.

REMARK. From $(*)$, the condition $C_1 \cap E = C_2 \cap E$ is numerically equivalent to

$$C_1 \in V_{(d,d-1),\delta_1,\backslash \pi',\pi''} \quad , \quad C_2 \in V_{d-1,\delta_2,/\pi',\pi''}$$

for some π', π'' such that $\pi' + \pi'' = \pi$.

REMARK. Setting $S(\pi) = \sum \ell_i$ for $\pi = [\ell_1, \dots]$, we have $S(\pi'') = \delta_1$ (follows from #conditions on $C_1 = \dim\{C_1\}$), $\delta_2 = \delta - d + 1 + S(\pi')$.

Now we have ℓ_1' fixed, mult. 1 points on E_1 that C_1 must contain. Let $\delta_1 - j$ of these lie on rulings contained in C_1. The j remaining rulings will pass through some of the $d + 1$ interior points. And $j \leq \ell_1''$. The number of choices is $\sum_{j=0}^{\ell_1''} \binom{\ell_1'}{\delta_1 - j} \binom{d+1}{j}$.

The remaining part of C_1, that is $C_{1,0}$, is smooth. The remaining choice: degree of variety of divisors of type $\pi''' \equiv [\ell_1'' - j, \ell_2'', \dots]$ on E, equal to $m(\pi''')n(\pi''')$, with $n(\pi)$ defined as $\frac{S(\pi)!}{\ell_1! \cdots \ell_r!} = \frac{(\sum \ell_i)!}{\prod(\ell_i!)}$.

Putting everything together:

$$N_{d,\delta} = \sum_{|\pi|=d-1} m(\pi) \cdot$$

$$\sum_{0 \leq \pi'' = [\ell_i''] \leq \pi, \pi' = \pi - \pi''} N_{d-1, \delta - d + 1 + S(\pi'), \backslash \pi'', \pi'} \cdot \sum_{j}^{\ell_1''} m(\pi''')n(\pi''') \binom{d+1}{j} \binom{\ell_1'}{S(\pi'') - j}$$

Along the same lines one can obtain true recursion formulas.

REMARKS. On the right-hand-side there is precisely one term corresponding to C_2 having δ nodes: this corresponds to $N_{d-1,\delta,0,d-1}$, with the same leading term as $N_{d-1,\delta}$. One can then derive estimates for $N_{d,\delta} - N_{d-1,\delta}$.

Fix δ, consider d as a variable. Claim: as functions of d, $N_{d,\delta}$ is a polynomial of degree 2δ. One finds

$$N_{d,\delta} - N_{d-1,\delta} \sim dN_{d-1,\delta-1} \sim d^{2\delta-1}$$

There are exactly two further terms contributing to $d^{2\delta-1}$: for C_1 with one ruling, $\pi = [d-1]$; and for C_1 smooth, simply tangent to E at a unique point, $\pi = [d-3, 1]$.

The leading and next-to-leading coefficients of $N_{d,\delta}$ in d have been computed by Y. Choi [Choi]:

$$a_{2\delta}^\delta = \frac{3^\delta}{\delta!} \quad , \quad a_{2\delta-1}^\delta = \frac{-2 \cdot 3^\delta}{(\delta-1)!}$$

Finally, note that rational curves are the hardest to treat from this approach.

4. Results and conjectures on the tautological ring of $\mathcal{M}_g$—C. Faber
October 3, 1996

First, recall Witten's conjecture=Kontsevich's theorem. Consider $\overline{M}_{g,n}$ and its universal family $\overline{C}_{g,n} = \overline{M}_{g,n+1} \to \overline{M}_{g,n}$. Let ω be the relative dualizing sheaf of this map, and denote the natural sections by σ_i, $i = 1, \dots, n$. We obtain n line bundles $\mathcal{L}_i = \sigma_i^* \omega$.

Witten's conjecture is about the intersection numbers of the $\mathcal{L}_i$:

DEFINITION. $\langle \tau_{d_1} \cdots \tau_{d_k} \rangle := c_1(\mathcal{L}_1)^{d_1} \cdots c_1(\mathcal{L}_k)^{d_k}$ on $\overline{M}_{g,k}$ if $\sum d_j = 3g - 3 + k$, and 0 otherwise. (Note: so this carries the genus information.)

The conjecture gives a complete recipe to compute these numbers. The ingredients are

(1) the string equation: $\left\langle \tau_0 \prod_{i=1}^{k} \tau_{d_i} \right\rangle = \sum_{i:\, d_i \geq 1} \langle \tau_{d_1} \cdots \tau_{d_i-1} \cdots \tau_{d_k} \rangle$;

(2) the dilaton equation: $\left\langle \tau_1 \prod_{i=1}^{k} \tau_{d_i} \right\rangle = (2g - 2 + k) \left\langle \prod_{i=1}^{k} \tau_{d_i} \right\rangle$;

(3) a recursion (KdV-equation): let $T = \prod_{j=0}^{m} \tau_j^{e_j}$; then for $n > 0$

$$(2n + 1) \left\langle \tau_n \tau_0^2 T \right\rangle = \frac{1}{4} \left\langle \tau_{n-1} \tau_0^4 T \right\rangle + \sum_{0 \leq a_j \leq e_j} \prod_{j=0}^{m} \binom{e_j}{a_j}$$

$$\cdot \left(\langle \tau_{n-1} \tau_0 T_1 \rangle \langle \tau_0^3 T_2 \rangle + 2 \langle \tau_{n-1} \tau_0^2 T_1 \rangle \langle \tau_0^2 T_2 \rangle \right)$$

where $T_1 = \prod_{j=0}^{m} \tau_j^{a_j}$, and $T = T_1 T_2$.

Witten showed that (1) and (2) hold; Kontsevich proved (3). (See [Witten], [Kontsevich2].)

For example, this allows you to compute the intersection numbers of Mumford's classes κ_i on $\overline{M}_g$. For the universal family $\overline{C}_g \xrightarrow{\pi} \overline{M}_g$, with relative dualizing sheaf ω, let $K = c_1(\omega)$ and $\kappa_i = \pi_*(K^{i+1}) \in A^i \overline{M}_g$ (note: all Chow rings are taken with $\mathbb{Q}$-coefficients); here is a recipe for the intersection numbers of the κ_i's: put

$$\langle \tau_{d_1+1} \tau_{d_2+1} \cdots \tau_{d_k+1} \rangle = \sum_{\sigma \in \Sigma_k} \kappa_\sigma$$

for $\sum d_i = 3g - 3$, where Σ_k denotes the symmetric group, and κ_σ is defined as follows: think of Σ_k as acting on the k-tuple $(d_1, \ldots, d_k)$; write σ as a product of disjoint cycles $\alpha_1 \alpha_2 \cdots \alpha_{\nu(\sigma)}$ (including 1-cycles). Then

$$\kappa_\sigma = \kappa_{|\alpha_1|} \kappa_{|\alpha_2|} \cdots \kappa_{|\alpha_{\nu(\sigma)}|}$$

where $|\alpha|$ = sum of the elements in the cycle α.

EXAMPLES. $k = 1$: $\langle \tau_{3g-2} \rangle = \kappa_{3g-3}$;
$k = 2$: $a + b = 3g - 3$, $\langle \tau_{a+1} \tau_{b+1} \rangle = \kappa_a \kappa_b + \kappa_{3g-3}$;
$k = 3$: $a + b + c = 3g - 3$, $\langle \tau_{a+1} \tau_{b+1} \tau_{c+1} \rangle = \kappa_a \kappa_b \kappa_c + \kappa_{a+b} \kappa_c + \cdots + 2\kappa_{3g-3}$.

(Alternative formulation, due to Zagier: set $\sigma_{d_1} \cdots \sigma_{d_k} = \langle \tau_{d_1+1} \cdots \tau_{d_k+1} \rangle$; then

$$\sigma_{abc\ldots x} = \sigma_{abc\ldots} \kappa_x + \sigma_{a+x,bc\ldots} + \sigma_{a,b+x,c\ldots} + \cdots$$

allows us to translate from σ's to κ's.)

Next, let's move to M_g by restricting the classes defined above.

THEOREM. $\kappa_{g-2} \neq 0$ in $A^{g-2}(\mathcal{M}_g)$.

This result should be compared with what was known before:

$\kappa_0 = 2g - 2 \neq 0$ for $g = 2$;

$\kappa_1 \neq 0$ for $g = 3$: M_3 contains complete curves, κ_1 is ample on M_3;

$\kappa_2 \neq 0$ for $g = 4$ (Faber's thesis, a long calculation). A simple proof can be obtained given that κ_1^2 and κ_2 are proportional on M_4. On $\overline{M}_g$ we have $\kappa_1 = 12\lambda_1 - \delta$ (δ = sum of boundary divisors). Showing $\lambda_1^2 \neq 0$ on M_4 is equivalent to showing $\lambda_1^2\lambda_3\lambda_4 \neq 0$ on $\overline{M}_4$ (as we will see in the first part of the proof below). Now map $\overline{M}_4$ to $\mathcal{A}_4^*$ (p.p.a.v.) by $C \to \mathrm{Jac}C$; the λ_i's are pull-backs from $\mathcal{A}_4^*$, and here $\lambda_1^2\lambda_3\lambda_4 \sim \lambda_1^9 \neq 0$ since a multiple of λ_1 embeds $\mathcal{A}_g^*$ in its Satake compactification.

Now back to the Theorem:

PROOF (SKETCH). Denote by $\mathbb{E}$ the locally free rank-g sheaf $\pi_*(\omega)$ on $\overline{M}_g$ (that is, the Hodge bundle). Then observe that, with $\lambda_i := c_i(\mathbb{E})$, $\lambda_g\lambda_{g-1}$ vanishes on $\overline{M}_g - M_g$. Indeed, $\overline{M}_g - M_g = \cup_{i=0}^{[g/2]}\Delta_i$, with $\Delta_0 =$ closure of the locus of nodal genus-$(g-1)$ curves and, for $i \geq 1$, $\Delta_i =$ closure of curves consisting of the union of a genus-i and a genus-$(g-i)$ curve. On (a finite cover of) Δ_0, we have the exact sequence

$$0 \to \mathbb{E}_{g-1} \to \mathbb{E}_g \to \mathcal{O} \to 0 \quad ,$$

so $\lambda_g = 0$ on Δ_0; on Δ_i, for $i > 0$:

$$\mathbb{E}_g = \mathbb{E}_i \oplus \mathbb{E}_{g-i}$$

and therefore $\lambda_g = (pr_i^*\lambda_i)(pr_{g-i}^*\lambda_{g-i})$, $\lambda_{g-1} = \dots$; and $\lambda_g\lambda_{g-1} = 0$ because in every genus h, $\lambda_h^2 = 0$. And why is $\lambda_h^2 = 0$? Mumford shows that $c(\mathbb{E})^{-1} = c(\mathbb{E}^\vee)$, which implies it right away.

(Alternate argument: $\forall k \geq 1$, $\mathrm{ch}_{2k}(\mathbb{E}) = 0$; hence $\forall \ell \geq 2g$, $\mathrm{ch}_\ell(\mathbb{E}) = 0$; $\mathrm{ch}_{2g-1}(\mathbb{E}) = (\text{nonzero } \#)\lambda_g\lambda_{g-1}$ must then vanish on $\overline{M}_g - M_g$ since the components here all have genus $< g$.)

The conclusion is that $\kappa_{g-2} \neq 0$ on $M_g \iff \kappa_{g-2}\lambda_{g-1}\lambda_g \neq 0$ on $\overline{M}_g$.

Now we need Mumford's expression for $\mathrm{ch}(\mathbb{E})$ obtained in [Mumford]:

$$\mathrm{ch}_{2g-1}(\mathbb{E}) = (\neq 0)\left[\kappa_{2g-1} + \frac{1}{2}\sum_{h=0}^{g-1}(i_h)_*(K_1^{2g-2} - K_1^{2g-3}K_2 + \cdots + K_2^{2g-2})\right]$$

Here $i_0 : \overline{M}_{g-1,2} \to \Delta_0 \subset \overline{M}_g$, and $K_i =$cotangent at the ith point; for $h > 0$, $i_h : \overline{M}_{h,1} \times \overline{M}_{g-h,1} \to \Delta_h \subset \overline{M}_g$, and K_1, K_2 are pull-backs from the factors.

With this,

$$\frac{1}{(\neq 0)}\kappa_{g-2}\mathrm{ch}_{2g-1}(\mathbb{E}) = \langle\tau_{g-1}\tau_{2g}\rangle - \langle\tau_{3g-2}\rangle + \frac{1}{2}\sum_{j=0}^{2g-2}(-1)^j\langle\tau_{2g-2-j}\tau_j\tau_{g-1}\rangle$$

$$+\frac{1}{2}\sum_{h=1}^{g-1}\left((-1)^{g-h}\langle\tau_{3h-g}\tau_{g-1}\rangle\langle\tau_{3(g-h)-2}\rangle + (-1)^h\langle\tau_{3h-2}\rangle\langle\tau_{3(g-h)-g}\tau_{g-1}\rangle\right)$$

$$=\frac{g!}{2^{g-1}(2g)!}$$

using the recurrence relation for the τ's in Witten's conjecture.

This implies $\kappa_{g-2}\lambda_{g-1}\lambda_g \neq 0$ on $\overline{M}_g$ and concludes the proof. $\square$

REMARK. The last, computational step is still rather complicated; it requires knowing certain 'n-point functions' explicitly for $n = 3$ (the n-point functions are $\sum\langle\tau_{d_1}\cdots\tau_{d_n}\rangle x_1^{d_1}\cdots x_n^{d_n}$). There is a nice formula for the 2-point function, due to Dijkgraaf: with $\tau(w) = \sum_{n\geq 0}\tau_n w^n$,

$$\langle\tau_0\tau(w)\tau(z)\rangle = \exp\left(\frac{w^3 + z^3}{24}\right)\sum_{n\geq 0}\frac{n!}{(2n+1)!}\left[\frac{1}{2}wz(w+z)\right]^n .$$

In the proof, I make use of an explicit formula that I found for the special 3-point function $\langle\tau(-w)\tau(w)\tau(z)\rangle$. (Recently, Zagier found such a formula for the general 3-point function.) Details of the proof can be found in [Faber2].

Next, we move to a result of Looijenga [Looijenga].

Let $\mathcal{C}_g^n \xrightarrow{\pi} M_g$ be the n-fold fiber product of $\mathcal{C}_g(= M_{g,1})$ over M_g, and denote by pr_i the projections onto the factors. We let $R^*(\mathcal{C}_g^n)$ be the *tautological ring* of $\mathcal{C}_g^n$:

for $n = 0$ $\mathcal{C}_g^0 = M_g$, and $R^*(M_g) :=$ the subring of $A^*(M_g)$ generated by the κ_i (restricted to M_g);

for $n \geq 1$, $R^*(\mathcal{C}_g^n) :=$ the subring of $A^*(\mathcal{C}_g^n)$ generated by the $\pi^*\kappa_i$, and the divisor classes $K_i = pr_i^*K$ and $D_{ij} = [\{x_i = x_j\}]$ (thinking of $\mathcal{C}_g^n$ as parametrizing objects $(C; x_1, \ldots, x_n)$).

THEOREM. *(Looijenga) A degree-d element in $R^*(\mathcal{C}_g^n)$ is a linear combination of classes of the irreducible components of*

$$\{(C; x_1, \ldots, x_n) \text{ such that there is an } f : C \to \mathbb{P}^1 \text{ of degree} \leq 2g - 2 + n, \text{ with }$$
$$\#f^{-1}(0) \leq g + n - 1 - d, \#f^{-1}(\infty) = 1, \{x_1, \ldots, x_n\} \subset f^{-1}(0) \cup f^{-1}(\infty)\}$$

Further, for $d = g + n - 2$, all the classes of these irreducible components are proportional to the class of

$$\mathcal{H}_g^n = \{(C; x_1, \ldots, x_n) \text{ such that } C \text{ is hyperelliptic,}$$
$$\text{and } x_1 = \cdots = x_n \text{ is a Weierstrass point of } C\}$$

COROLLARY. *For $d > g + n - 2$, $R^d(C_g^n) = 0$.*

COROLLARY. *For $d = g + n - 2$, $R^d(C_g^n)$ is at most one-dimensional.*

The first Theorem given above, together with this last result, imply that in fact $R^{g+n-2}(C_g^n) = \mathbb{Q}$.

PROOF. (Sketch of the first part.)

(1) It is enough to prove the statement for the monomials in the K_i only. We have then to prove a statement as in the theorem, but with $\deg f \leq g + n$.

(2) Simple observation: if $D_0 \sim D_\infty$ are positive disjoint divisors on a curve C, then there is a $\pi : C \to \mathbb{P}^1$ such that $\pi^*(i) = D_i$, $i = 0, \infty$; and if $p \in C$ occurs in D_i with multiplicity m_p, π induces a map $\mathbb{C} = T_i^* \mathbb{P}^1 \xrightarrow{\pi^*} T_p^* C^{\otimes m_p}$. This is not canonical because π is not canonical. But let $R =$ ramification outside 0 and ∞, and consider $\pi_* R$; fix π so that $\prod_{z \in \pi_* R} z = 1$: this determines π up to a $(\deg R)$-th root of unity.

(3) (Lemma 2.4 in loc.cit.) Relativize over a disk with generic point η and closed point 0. Consider $C \to \Delta$ with section $\Delta \xrightarrow{x} C$. Let $\mathcal{P}$ be a relative pencil with $d(x)$ as a member, and assume that $\mathcal{P}_\eta$ is base-point-free. $C_\eta \to \mathbb{P}^1_\eta$ is ramified at R_η outside x_η. Specialize R_η to R_0, obtain:

$\text{mult}_{x(0)} R_0 =$ mult. of $x(0)$ in the fixed part of $\mathcal{P}_0$.

$\implies$ (4) $\forall$ member $D \neq d(x)$ of $\mathcal{P}$ specializing to $d(x(0))$, the degree of the moving part of $\mathcal{P}_0$ is $\leq$ the number of η-valued points of $\{\text{supp}(D_\eta) - x(\eta)\}$

(5) Let Z be the moduli space of tuples $(C; x_1, \ldots, x_n, x; D; \mathcal{P})$ with $x_i, x \in C$; $\mathcal{P}$ a pencil on C, with $(n + g)x$ as a member; and D a degenerate member of the pencil (so $\#\text{supp}(D) < n + g$) with $\{x_1, \ldots, x_n\} \subset \text{supp}(D)$.

Stratify Z by:

$$Z^k = \{\text{supp}(D) \text{ has } \leq g + n - 1 - k \text{ points outside } x\}; \text{ so that}$$

$$Z^{n+g-1} = \{(C; \underbrace{x, \ldots, x}_{n}, x; (n + g)x, \mathcal{P}\}$$

LEMMA. *$\forall k < g + n - 1$: $Z^k - Z^{k-1}$ is quasi-affine of pure dimension $(3g - 3 + n - k)$; and $f^* K_i = 0$ on $Z^k - Z^{k+1}$ for $i = 1, \ldots, n$.*

(6) Define X^k to be the union of the irreducible components of Z^k that are not contained in Z^{n+g-1}.

(a) $f : X^0 \to C_g^n$ is proper and surjective.

(b) *Claim 1:* $f(X^k \cap Z^{n+g-1}) \subset f(X^{k+1})$

Namely, for $A \in X^k \cap Z^{n+g-1}$, write $A = (C; x, \ldots, x, x; (n+g)x, P)$, so that $f(A) = (C; x, \ldots, x)$; by (4), the moving part of P has degree $\leq n + g - k - 1$ so that P has a member $E \neq (n + g)x$ with $\leq n + g - k - 2$ points outside x. Then $B = (C; x, \ldots, x, x; E, P) \in Z^{k+1}$, $\notin Z^{n+g-1}$, so $\in X^{k+1}$; $f(A) = f(B)$, proving the claim.

Now set $U^k = f^{-1}(f(X^k) - f(X^{k+1})) \subset X^k$

Claim 2: $U^k \subset Z^k - Z^{k+1}$.

This is easy to see: $U^k \cap Z^{n+g-1} = \emptyset$ (because $a \in U^k \cap Z^{n+g-1} \implies f(a) \in f(X^{k+1})$, contradiction); hence $U^k \cap Z^{k+1} = \emptyset$ (because $a \in U^k \cap Z^{k+1} \implies a \notin Z^{n+g-1} \implies a \in X^{k+1}$, contradiction).

It follows that $f^* K_i = 0$ on U^k; $f : U^k \to f(X^k) - f(X^{k+1})$ is proper, onto, finite, hence $K_i = 0$ on $f(X^k) - f(X^{k+1})$ (using $\mathbb{Q}$-coefficients).

From this: all monomials of degree d in the K_i are supported on $f(X^d)$; by (1), this is enough to prove the first part of the theorem. $\square$

The results seen so far support a standing conjecture on the tautological ring of M_g:

CONJECTURE. (1) $R^*(M_g)$ is Gorenstein with socle in degree $(g-2)$. That is:

(i) $R^j(M_g) = 0$ for $j > g - 2$;
(ii) $R^{g-2}(M_g) \cong \mathbb{Q}$;
(iii) there is a perfect pairing $R^i(M_g) \times R^{g-2-i}(M_g) \to R^{g-2}(M_g) \overset{\text{fix}}{=} \mathbb{Q}$.

(2) $\kappa_1, \ldots, \kappa_{[g/3]}$ generate the ring, no relations up to degree $[g/3]$.
(3) Explicit proportionality factors in degree $(g-2)$.

As we have seen, parts (i) and (ii) of (1) are now proved. The rest of the conjecture is still open, although we have been able to check it for all $g \leq 15$ (and we have reduced it to a hard combinatorial problem for all g). A complete statement, and discussion of the evidence for this conjecture, can be found in [Faber].

---

5. Counting rational curves on quintic 3-folds—S. Kleiman
October 10, 1996

This talk is a report on joint work with T. Johnsen, which was presented in two papers, [J-K] and alg-geom/9601024. The aim here is to place this work in context, to explain the main results, and to give the flavor of the proofs. The talk is organized into these three sections:

I. Context
II. Strategy
III. Proofs

I. Context. Let F be a hypersurface in $\mathbb{P}^4$ of degree 5 over the complex numbers. Assume that F is *general*, that is, represented by a point in a suitable Zariski-open set of $\mathbb{P}^{125}$. (By contrast, F is called *generic*, if it's represented by a point in the intersection of countably many Zariski-open sets. The latter condition is necessary when we consider all degrees simultaneously, but here we will consider only small degrees.)

Every irreducible rational curve C of degree d in $\mathbb{P}^4$ is given by a parameterization of the form,

$$\big(\alpha_0(t, u), \ldots, \alpha_4(t, u)\big),$$

where α_i is a homogeneous polynomial of degree d. The smooth C form an open set in $\mathrm{Hilb}(\mathbb{P}^4)$; denote it by M_d. Each parameterization is represented by a point of an affine space, and those parameterizations giving smooth curves of degree d form an open subset, which maps onto M_d; hence M_d is irreducible. This affine space has dimension $5(d + 1)$, and the fiber over a C in M_d has dimension 4; hence,

$$\dim M_d = 5d + 1.$$

In fact, it's not hard to compute the dimensions of the cohomology groups of the normal bundle $\mathcal{N}_C\mathbb{P}^4$; whence, by the standard infinitesimal theory of the Hilbert scheme, M_d is smooth of dimension $5d + 1$.

An arbitrary C of degree d lies on F if and only if the polynomial in t, u,

$$F(\alpha_0(t, u), \ldots, \alpha_4(t, u)),$$

is identically zero. We expect this polynomial to be homogeneous of degree $5d$, and have $5d+1$ coefficients. Their vanishing would impose $5d+1$ conditions on C. So, since $\dim M_d = 5d + 1$, we expect only finitely many C on F. Denote the number of smooth C on F by n_d, and the number of all C on F by n'_d.

Schubert (1885): The number of lines on F is $n_1 = 2875$.

Clemens (1983, '84, '86): After having proved that the Griffiths group of F has infinite rational rank (that is, the vector space $(G_h(F)/G_a(F)) \otimes \mathbb{Q}$ is infinite dimensional) when F is generic, Clemens made the following series of conjectures about the irreducible rational curves C of degree d on F:

   (a) $1 \leq n_d < \infty$.
   (b) Each smooth C is infinitesimally rigid on F.
   (c) There are no singular C on F, and so $n_d = n'_d$.
   (d) Any two C, C' are disjoint.
   (e) $n_d = 5^3 \cdot d \cdot *$.

S. Katz (1986): Conjectures (a) and (b) hold for $d \leq 7$. The number of conics on F is, $n_2 = 609,250$, which is of the form $5^3 \cdot 2 \cdot *$ prescribed by (e).

   THEOREM 1 (Katz for $d \leq 7$, Nijsse and Johnsen–Kleiman for $d = 8,9$). *Conjectures* (a) *and* (b) *hold for* $d \leq 9$.

Vainsencher (1995): Conjecture (c) is false. In fact, there are $17,601,000$ six-nodal plane quintic curves on F, arising from tangent 2-planes. These curves deform to smooth irreducible curves (so they are not infinitesimally rigid); however, the corresponding maps $\mathbb{P}^1 \to F$ are rigid.

   PROPOSITION 2 (Johnsen–Kleiman). *There are no 16-nodal 10-ics on* F *arising from tangent quadric surfaces.*

   THEOREM 3 (Johnsen–Kleiman). *In degree* d *at most 9, the following variations of Conjectures* (c) *and* (d) *hold:*

   (c') *There are no singular* C *on* F, *other than Vainsencher's quintics.*
   (d') *Any two* C, C' *are disjoint if* $\deg C + \deg C' \leq 9$ *(including Vainsencher's quintics).*

Ellingsrud–Strømme (1991, '93): The number of twisted cubics on F is $n_3 = 371,206,375$. Note that 3 does not divide n_3, disproving part of Conjecture (e).

Candelas–de la Ossa–Green–Parkes (1990): Mirror symmetry gives an algorithm for finding a suitably defined number n'_d for all d. The values for $d \leq 10$ are given, and 5^3 *does* divide n'_d for $d \leq 10$.

Lian–Yau (1994): If 5 does not divide d, then 5^3 divides n'_d, as defined in [CDGP]. It's not ruled out that 5^3 always divides n'_d.

Kontsevich (1994): Set $N_d := c_{\text{top}} E_d$, where E_d is the vector bundle on $\overline{M}_{0,0}(\mathbb{P}^4, d)$ obtained as $\phi_* \rho^* \mathcal{O}(5)$ via the diagram,

$$\overline{M}_{0,1}(\mathbb{P}^4, d) \xrightarrow{\ \rho\ } \mathbb{P}^4$$
$$\downarrow{\phi}$$
$$\overline{M}_{0,0}(\mathbb{P}^4, d)$$

Kontsevich computed N_4, and it leads to the same n'_4 computed in [CDGP].

Note that there's a positive-dimensional locus in the zero set of the section of E_d defined by F, namely,

$$\{\text{maps } \mu \colon \mathbb{P}^1 \to C \subset F \text{ of degree } k\}.$$

It has dimension $(2k - 2)$. To get from N_d to n'_d, use residual intersection theory, which yields the formula,

$$N_d = \sum_{k|d} \frac{n'_{d/k}}{k^3}.$$

This number n'_d is, however, not yet proved to be equal to the one computed by the physicists in [CDGP]; however, it *is* equal to the number of irreducible C on F if the latter number is finite.

II. Strategy. In two words, the strategy (due to Clemens and Katz) is this: count constants. Namely, form the incidence variety,

$$I_d := \{(C, F) | C \subset F\} \subset M_d \times \mathbb{P}^{125}.$$

The naive count of parameterizations above shows that every component of I_d has dimension at least 125.

THEOREM 4 (Clemens–Katz). *For all d, there is a pair (C, F) in I_d with F smooth along C, and with C smooth and infinitesimally rigid; in fact,*

$$\mathcal{N}_C F = \mathcal{O}_{\mathbb{P}^1}(-1) \oplus \mathcal{O}_{\mathbb{P}^1}(-1).$$

COROLLARY 5 (Katz). *If I_d is irreducible for a given d, then Conjectures (a) and (b) hold for this d.*

PROOF. The projection $I_d \to \mathbb{P}^{125}$ is smooth and finite at a *Clemens–Katz pair* (C, F), that is, a pair given by the theorem. Therefore, the projection is smooth and finite over a generic F if I_d is irreducible. □

THEOREM 6 (Katz for $d \leq 7$, Nijsse and Johnsen–Kleiman for $d = 8, 9$). *The incidence variety I_d is irreducible for $d \leq 9$.*

COROLLARY 7 (Johnsen–Kleiman). *A smooth C on F of degree d at most 9 has the following properties:*

(1) *C spans a d-plane if $d \leq 4$; otherwise, it spans $\mathbb{P}^4$.*
(2) *C is of maximal rank; that is, for each k, the restriction map,*

$$\rho_k \colon H^0(\mathcal{O}_{\mathbb{P}^4}(k)) \to H^0(\mathcal{O}_C(k)),$$

is either injective or surjective or both.
(3) *For $d = 4q + r$ with $0 \leq r < 4$, the restricted twisted sheaf of 1-forms decomposes as follows:*

$$\Omega^1_{\mathbb{P}^4}(1)|C = \mathcal{O}_{\mathbb{P}^1}(-q-1)^r \oplus \mathcal{O}_{\mathbb{P}^1}(-q)^{4-r}.$$

PROOF. The projection $I_d \to M_d$ is surjective for $d \leq 9$. Properties (1)–(3) hold for a general C by direct computation and by results of Ballico–Ellia and Verdier. □

PROPOSITION 8 (Johnsen–Kleiman). *The incidence variety I_d has the following properties:*

(1) *If $d \geq 25$, then the projection $I_d \to M_d$ is not surjective.*
(2) *If $d \geq 12$, then I_d is reducible.*
(3) *If $d \leq 24$, then there exists a unique component of I_d covering M_d, and it has dimension 125.*

PROOF. (1) Since $5d + 1 \geq 126$ for $d \geq 25$, the restriction map ρ_5 is not surjective. Now, the maximal rank theorem says that a general C in M_d is of maximal rank. Hence, ρ_5 is injective for a general C, and so $H^0(\mathcal{I}_C(5)) = 0$ where $\mathcal{I}_C$ is the ideal of C in $\mathbb{P}^4$. Therefore, a general C lies in no F.

(2) Consider the locus of pairs (C, F) in I_d such that C lies on a smooth quadric surface. This locus has dimension $2d + 101$. On the other hand, a Clemens–Katz pair lies in a (unique) component of I_d of dimension 125.

(3) The maximal rank theorem implies that the locus,

$$I_{d,0} := \{(C, F) \in I_d | H^1(\mathcal{I}_C(5)) = 0\},$$

is nonempty if and only if $d \leq 24$. Its closure $\overline{I_{d,0}}$ is the component in question. More details can be found at the beginning of the next section. □

CONJECTURE 9 (Johnsen–Kleiman). *For $d \leq 24$, the complement $I'_d :=$ $I_d - I_{d,0}$ does not cover $\mathbb{P}^{125}$.*

PROPOSITION 10 (Johnsen–Kleiman). *For $d \leq 24$, Conjecture 9 implies Conjectures* (a), (b) *and Corollary 7.*

CONJECTURE 11 (Johnsen–Kleiman). *For $d \leq 24$, the component $\overline{I_{d,0}}$ of I_d contains all the Clemens–Katz pairs (C, F).*

III. Proofs. Consider the projection $\alpha \colon I_d \to M_d$, and let $C \in M_d$. The fiber $\alpha^{-1}C$ is equal to $\mathbb{P}(H^0(\mathcal{I}_C(5)))$. To compute its dimension, form the sequence,

$$0 \to \mathcal{I}_C \to \mathcal{O}_{\mathbb{P}^4} \to \mathcal{O}_C \to 0.$$

The associated long exact sequence,

$$0 \to H^0(\mathcal{I}_C(5)) \to H^0(\mathcal{O}_{\mathbb{P}^4}(5)) \xrightarrow{\rho_5} H^0(\mathcal{O}_C(5)) \to H^1(\mathcal{I}_C(5)) \to H^1(\mathcal{O}_{\mathbb{P}^4}(5)),$$

in which the last term vanishes, yields the formulas,

$$\dim \alpha^{-1}C = h^0(\mathcal{I}_C(5)) - 1 = 124 - 5d + h^1(\mathcal{I}_C(5)).$$

Set $M_{d,i} := \{C \in M_d | h^1(\mathcal{I}_C(5)) = i\}$ and $I_{d,i} := \alpha^{-1}M_{d,i}$. Then $M_{d,0}$ is open, and the maximal rank theorem implies that, if $d \leq 24$, then $M_{d,0}$ is nonempty. Recall that M_d is irreducible. Therefore, if $d \leq 24$, then $I_{d,0}$ is irreducible of dimension 125, and its closure is the component in question in Part (3) of Proposition 8.

By a theorem of Gruson, Lazarsfeld, and Peskine, if $d \leq 7$, then $M_{d,0} = M_d$, and so I_d is irreducible. For $d = 8, 9$, we have to work a little harder, and prove the following lemma. It implies that $I'_d := I_d - I_{d,0}$ has dimension at most 124. Since every component of I_d must have dimension at least 125, again I_d is irreducible. Thus Theorem 6 holds, and Theorem 1 follows because of Corollary 5.

LEMMA 12. *For $d = 8, 9$, if $i \geq 1$, then* $\mathrm{codim}(M_{d,i}, M_d) \geq i + 1$.

PROOF. Assume $d = 8$. Then $\dim M_8 = 5(8 + 1) - 4 = 41$. Now, the work of Gruson, Lazarsfeld, and Peskine tells us that

$$h^1(\mathcal{I}_C(5)) = \begin{cases} 0, & \text{if } C \text{ spans } \mathbb{P}^4; \\ 1, & \text{if } C \subset \mathbb{P}^3 \text{ and } C \not\subset \text{ a smooth quadric } Q; \\ 5, & \text{if } C \subset \text{ a smooth quadric } Q \subset \mathbb{P}^3. \end{cases}$$

Hence $\dim M_{8,1} = 4(8+1) - 4 + 4 = 36$: the first term in the middle is the number of parameterized C in $\mathbb{P}^3$; the second is the number of reparameterizations, and the third is the number of $\mathbb{P}^3$s in $\mathbb{P}^4$. Similarly, we have

$$\dim M_{8,5} = \dim\{C \subset Q\} + \dim\{Q \subset \mathbb{P}^3\} + \dim\{\mathbb{P}^3 \subset \mathbb{P}^4\}$$
$$= 15 + 9 + 4 = 28.$$

Assume $d = 9$. By the work of Gruson, Lazarsfeld, and Peskine again and by an extension of it due to d'Almeida, there are five cases to consider:

$$h^1(\mathcal{I}_C(5)) \leq \begin{cases} 0, & C \not\subset \text{hyperplane, has no 7-secants;} \\ 0, & C \subset \text{hyperplane, has no 7-secants;} \\ 1, & C \not\subset \text{hyperplane, has 7-secants;} \\ 10, & C \subset \text{smooth quadric surface (so has 7-secants);} \\ 8, & C \subset \text{hyperplane, } \not\subset \text{smooth quadric, has 7-secants.} \end{cases}$$

In fact, equality holds in the first four cases, and a certain amount of direct analysis is required to handle the last two cases. Given these bounds, the assertion is established by counting the number of C that appear in each of the five cases. □

The proof of Theorem 3 is, in spirit, like that of Theorem 6. Namely, we decompose a suitable incidence variety of pairs (C, F) into manageable locally closed pieces, whose dimensions we can bound from above using the work of Gruson, Lazarsfeld, and Peskine and of d'Almeida. We conclude that these pieces do not cover $\mathbb{P}^{125}$, and so a general F contains no C in question.

For (c′), the C in question are the *singular* irreducible rational curves of degree d at most 9. Since we only need to prove crude bounds on dimensions, we may work with pieces of the space of parameterized C rather than the Hilbert scheme, and we do so to simplify the job. On the other hand, each C has a nonzero arithmetic genus g, which enters the scene via the Riemann–Roch theorem,

$$h^0(\mathcal{O}_C(5)) = 5d + 1 - g + h^1(\mathcal{O}_C(5)).$$

So we must use the Castelnuovo–Halphen bounds on g in terms of d. There are many cases to analyze, and the analysis is at times a bit tedious.

For (d′), the C in question are the reducible curves of degree d at most 9 with two intersecting components A and B, which are either both smooth curves with the three properties listed in the corollary to Theorem 3 or else one is such a smooth curve and the other is a six-nodal plane quintic. The main ingredients in the proof are the following two lemmas. The first is proved via a direct case-by-case analysis, and the second is proved using Hirschowitz's "méthode d'Horace" in the case where one of the components is a six-nodal plane quintic.

LEMMA 13. *We have* $\operatorname{codim}\{C | \#(A \cap B) \geq n + 1\} \geq n$.

LEMMA 14. *We have* $h^1(\mathcal{I}_C(5)) = 0$.

6. Operads and associativity of QH^*—P. Aluffi
October 15, 1996

This lecture is prompted by a remark in §10 of [F-P], stating that the associativity of the quantum product is in a suitable sense equivalent to a certain map being a morphism of *operads*. I will set up the definitions necessary to understand this statement, and give a sketchy indication of why it holds. The plan is as follows:

—§1. *Example*
—§2. *Formal definition*
—§3. *The endomorphism operad of a vector space V, End_V*
—§4. *The moduli space operad, $\mathcal{M}$, and its homology, $H_*\mathcal{M}$*
—§5. *Gromov-Witten numbers and Gromov-Witten classes*
—§6. *GW on X induce $H_*\mathcal{M}(n) \to End_{H^*X}(n)$*
—§7. *Properties of GW and morphisms of operads*

§1. **Example.** A "disk arrangement" is a disjoint union of labeled disks within the unit circle:

Let $O(n)$ denote the set of all n-disk arrangements ($n \geq 0$). S_n acts on $O(n)$ by relabeling the subdisks. We can define an operation

$$\rho : O(k) \times O(\ell_1) \times \cdots \times O(\ell_k) \to O(\ell_1 + \cdots + \ell_k)$$

by the following recipe: given $(o(k); o(\ell_1), \ldots, o(\ell_k))$ in the source,
    —scale the unit circle in $o(\ell_i)$ down to the size of the ith subdisk in $o(k)$;
    —replace the ith subdisk of $o(k)$ with the resized $o(\ell_i)$;
    —remove the boundary of the resized $o(\ell_i)$;
    —label the total $\sum \ell_i$ subdisks in the natural way (starting from the first ℓ_1, etc.).
For example, $\rho : O(2) \times O(1) \times O(3) \to O(4)$ acts:

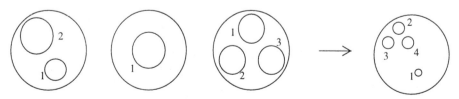

We should think of each $o(k) \in O(k)$ as giving a different 'multiplication' from $O(\ell_1) \times \cdots \times O(\ell_k)$ to $O(\ell_1 + \cdots + \ell_k)$.

This system of operations, for all k, satisfies an obvious "associativity" rule. Suppose given

—a disk arrangement $o(k)$;

—k disk arrangements $o(\ell_1), \ldots, o(\ell_k)$; and

—for each i, ℓ_i disk arrangements $o(m_{i1}), \ldots, o(m_{i\ell_i})$ (for a total of $\sum_{i,j} m_{ij}$ subdisks).

Then we can do two different things:

—first apply ρ to each $(o(\ell_i); o(m_{i1}), \ldots, o(m_{i\ell_i}))$, producing an $o(\sum_j m_{ij})$ for each i; then apply to $(o(k); o(\sum_j m_{1j}), \ldots, o(\sum_j m_{kj}))$; or

—first apply ρ to $(o(k); o(\ell_1), \ldots, o(\ell_k))$, obtaining an $o(\sum \ell_i)$; then apply to $(o(\sum \ell_i); o(m_{11}), \ldots, o(m_{k\ell_k}))$.

It is clear that these two operations produce the same $o(\sum_{ij} m_{ij})$. In other words, the following **operadic diagram** commutes:

$$O(k) \times (O(\ell_1) \times \textstyle\prod_j O(m_{1j})) \times \cdots \times (O(\ell_k) \times \prod_j O(m_{kj})) \xrightarrow{\; id \times \rho \times \cdots \times \rho \;} \cdots$$

$$\downarrow \text{shuffle}$$

$$(O(k) \times O(\ell_1) \times \cdots \times O(\ell_k)) \times O(m_{11}) \times \cdots \times O(m_{k\ell_i}) \xrightarrow{\; \rho \times id \times \cdots \times id \;} \cdots$$

$$\cdots \longrightarrow \quad O(k) \times O(\textstyle\sum_j m_{1j}) \times \cdots \times O(\sum_j m_{kj}) \xrightarrow{\; \rho \;} O(\sum_{ij} m_{ij})$$

$$\|$$

$$\cdots \longrightarrow O(\ell_1 + \cdots + \ell_k) \times O(m_{11}) \times \cdots \times O(m_{k\ell_k}) \xrightarrow{\; \rho \;} O(\textstyle\sum_{ij} m_{ij})$$

There is more structure in this simple example. In $O(1)$ there is a special Q, that is the 'unit disk inside the unit circle'. It's clear that this acts as a unit, for example in the sense that via $O(1) \times O(n) \to O(n)$, for all $o(n) \in O(n)$ we have $(Q, o(n)) \to o(n)$. To state this in the proper generality, we can say that *there exists a map q from* the singleton, that is *the unit element Q for product* in the category of sets, *to $O(1)$, such that the following two **unit diagrams** commute:*

$$
\begin{array}{ccc}
Q \times O(n) & \xrightarrow{\;\sim\;} & O(n) \\
{\scriptstyle q \times id}\downarrow & \nearrow & \\
O(1) \times O(n) & &
\end{array}
\qquad
\begin{array}{ccc}
O(n) \times Q^n & \xrightarrow{\;\sim\;} & O(n) \\
{\scriptstyle id \times q^n}\downarrow & \nearrow & \\
O(n) \times O(1)^n & &
\end{array}
$$

There is even more structure: there are two obvious equivariance properties satisfied by ρ under the action of the permutation groups on the $O(n)$. These can be expressed by two **equivariance diagrams**, which are conceptually elementary but notationally demanding. These are left to the reader to write out; or see [May], for example.

§2. Formal definition. Let S be a symmetric monoidal category (that is, S has an associative product with unit, and shuffling of factors gives canonical isomorphisms) with product $\times$ and unit object Q. Examples we will consider will be

—Sets with product $\times$ and $Q =$ singleton;
—Topological Spaces, with product $\times$ and $Q =$ singleton;
—Vector Spaces over a field k, with product $\otimes$ and $Q = k$.

DEFINITION. An operad O in S consists of objects $O(n)$ for all $n \geq 0$, a unit map $Q \xrightarrow{q} O(1)$, a right action of the symmetric group S_n on $O(n)$ for each n, and operations

$$\rho : O(k) \times O(\ell_1) \times \cdots \times O(\ell_k) \to O(\ell_1 + \cdots + \ell_k)$$

for $k \geq 1$, satisfying the operadic, unit, and equivariance diagrams of §1.

REMARK. $O(0)$ plays little rôle in this definition, and no rôle in this lecture, so we will ignore it here.

Morphisms of operads $C \to O$ are defined in the natural way.

MORE EXAMPLES. • $C_n(k) =$ affine embeddings of k disjoint copies of the standard cube I^n in I^n ($C_n(k)$ can be suitably topologized, making this a topological operad). This is a straightforward generalization of the first example, and in some sense it was the 'first' operad. It is called the *Boardman-Vogt little n-cubes operad* (see [B-V]. There are no 'operads' there, as the name had not yet entered into use; there are however 'PROPS', 'cherry trees', etc.). The name and formal definition of operad were introduced by J. P. May, [May2]. A brief (pre)historical sketch on operads is in [Stasheff].

• *Oriented trees.* $\mathcal{T}(n) =$ the set of trees with one root and n labeled tails. For example, $\mathcal{T}(4)$ consists of

etc.

The operation is by 'grafting'. For example,

$$\mathcal{T}(3) \times \mathcal{T}(3) \times \mathcal{T}(1) \times \mathcal{T}(2) \to \mathcal{T}(6) \qquad \text{acts}$$

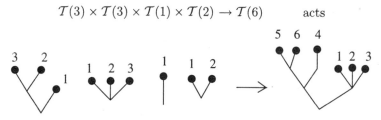

These are the 'cherry trees' of [B-V]. As graphs can be thought of as a generalization of trees, so there is a corresponding generalization of the notion of operads, that is *modular operads,* introduced by Getzler and Kapranov.

§3. **The Endomorphism operad.** Let V be a vector space (or more generally an element of a category as above and with internal Hom).

DEFINITION. The *Endomorphism operad* End$_V$ of V is defined by setting

$$\text{End}_V(n) = \text{Hom}(V^{\otimes n}, V) \quad .$$

The unit is the map $k \to \text{Hom}(V, V)$ sending 1 to the identity; the action of S_n is by permutation of the factors in $V^{\otimes n}$; and the operation

$$\text{End}_V(n) \otimes \text{End}_V(\ell_1) \otimes \cdots \otimes \text{End}_V(\ell_n) \to \text{End}_V(\ell_1 + \cdots + \ell_n)$$

acts on basic tensors as follows: for $(\phi \otimes \alpha_1 \otimes \cdots \otimes \alpha_n)$ in the source, that is $\phi \in \text{Hom}(V^{\otimes n}, V)$, $\alpha_i \in \text{Hom}(V^{\otimes \ell_i}, V)$, the corresponding homomorphism $V^{\otimes (\sum \ell_i)} \to V$ is induced by

$$(v_{11}, \ldots, v_{1\ell_1}, \ldots, v_{k1}, \ldots, v_{k\ell_k}) \mapsto \phi(\alpha_1(v_{11}, \ldots, v_{1\ell_1}), \ldots, \alpha_n(v_{n1}, \ldots, v_{n\ell_n}))$$

The operad axioms should be clear for End$_V$, from the associativity of composition.

Aside on terminology. If we have a morphism of operads $\mathcal{C} \to \text{End}_V$, we may say that '$V$ is a $\mathcal{C}$-algebra', or that we have defined an 'action of $\mathcal{C}$ on V', or that V is realized as a representation of $\mathcal{C}$ (again, these notions may be defined not just for vector spaces, but for objects in any reasonable category). May's original result was that "a connected space admits an action of the little cube operad $\mathcal{C}_n$ if and only if it has the homotopy type of an n-fold iterated loop space." There are a number of fancy-sounding terms that translate into "representation of (a certain) operad": see several papers by T. Kimura et al. For example, in [KSV] we read that " ... a conformal field theory at the tree level is equivalent to an algebra over the operad of Riemann spheres with punctures." Operads are increasingly relevant to physics, as are other fields that I usually would not associate with physics. In the same paper, I was surprised to read: "We recall the Deligne-Knudsen-Mumford compactification of $M_{0,n}$ ... see [11,23,24,25] or *any review of two-dimensional quantum field theory.*"

§4. **The moduli space operad.**

DEFINITION. The moduli space operad $\mathcal{M}$ is defined by setting $\mathcal{M}(n) = \overline{M}_{0,n+1}$ for $n \geq 2$, and $\mathcal{M}(1) =$ pt., to be pictured as a $\mathbb{P}^1$ with two marked points (that is, a component which will automatically contract to a point, by the stability requirement).

The unit of $\mathcal{M}$ consists of the single point in $\mathcal{M}(1)$; the action of S_n on $\mathcal{M}(n) = \overline{M}_{0,n+1}$ is by permutation of the first n points. For the operation

$$\rho : \mathcal{M}(k) \times \mathcal{M}(\ell_1) \times \cdots \times \mathcal{M}(\ell_k) \to \mathcal{M}(\ell_1 + \cdots + \ell_k) \quad ,$$

say that $(C(k); C(\ell_1), \ldots, C(\ell_k))$ is in the source (that is, $C(k)$ is a stable $(k+1)$-pointed rational curve, etc.) The operation joins $C(\ell_i)$ to $C(k)$ by identifying the last, 'free' marked point of $C(\ell_i)$ with the ith point of $C(k)$. After collapsing unstable components, this produces a stable curve with $(\ell_1 + \cdots + \ell_k) + 1$ marked points, as needed. A schematic representation of this operation on general points of the factors is:

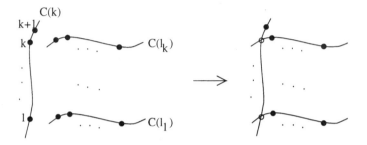

It is a good exercise to see what this operation does to the 'combinatorial type' of stable curves, in the sense of Belorousski's lecture on $\overline{M}_{0,n}$.

The sense in which $\mathcal{M}(1)$ acts as a unit is by collapsing the corresponding tail: for example, via $\mathcal{M}(1) \times \mathcal{M}(n) \to \mathcal{M}(n)$

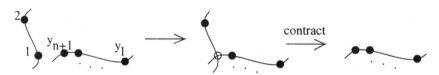

The operad $\mathcal{M}$ lives (for example) in the topological category. We can get a related operad in Vector Spaces over $\mathbb{Q}$ by taking homology: set

$$H_*\mathcal{M}(n) = H_*(\mathcal{M}(n), \mathbb{Q})$$

The operation is obtained by composing

$$H_*\mathcal{M}(k) \otimes H_*\mathcal{M}(\ell_1) \otimes \cdots \otimes H_*\mathcal{M}(\ell_k) \xrightarrow{\text{Künneth}} H_*(\mathcal{M}(k) \times \mathcal{M}(\ell_1) \times \cdots \times \mathcal{M}(\ell_k))$$

$$\xrightarrow{H_*(\rho)} H_*\mathcal{M}(\textstyle\sum \ell_i)$$

Next, the Gromov-Witten invariants on a variety X allow us to define linear maps $H_*\mathcal{M}(n) \to \text{End}_{H^*X}(n)$ for all $n \geq 1$. Then the properties of Gromov-Witten invariants will imply that this is a map of operads.

§5. **Gromov-Witten invariants and classes.** Reminder: for nice X and $\beta \in A^1 X$ there is a space $\overline{M}_{0,n}(X, \beta)$ and n evaluation maps $\eta_1, \ldots, \eta_n : \overline{M}_{0,n}(X, \beta) \to X$, acting $\eta_i : (C; p_1, \ldots, p_n; f) \mapsto f(p_i)$. The number associated with classes $\gamma_1, \ldots, \gamma_n \in A^*X$ is

$$\int_{\overline{M}_{0,n}(X,\beta)} \eta_1^* \gamma_1 \cup \cdots \cup \eta_n^* \gamma_n$$

Now we need to shift the focus a little: first, we want to consider the whole class $\eta_1^*\gamma_1 \cup \cdots \cup \eta_n^*\gamma_n$ (note: not gaining information, according to the 'first reconstruction theorem' in [K-M]); second, we transfer the class to $\overline{M}_{0,n}$. For this, consider the diagram

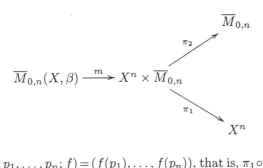

Here $(\pi_1 \circ m)(C; p_1, \ldots, p_n; f) = (f(p_1), \ldots, f(p_n))$, that is, $\pi_1 \circ m = (\eta_1, \ldots, \eta_n)$; we can then map $\gamma_1 \otimes \cdots \otimes \gamma_n$ in $H^*(X^n)$ to a class in $\overline{M}_{0,n}$ by pushing-forward via $(\pi_2 \circ m)$ the pull-back via $(\pi_1 \circ m)$.

DEFINITION. Call this class $I_{0,n,\beta}^X(\gamma_1 \otimes \cdots \otimes \gamma_n)$.

This definition reflects the *motivic axiom* of [K-M]. *Motivic* here means that the map $H^*X^n \to H^*\overline{M}_{0,n}$ is obtained via a correspondence, that is a class in $H_*(X^n \times \overline{M}_{0,n})$. For genus=0, and nice X, the image from $\overline{M}_{0,n}(X, \beta)$ of the fundamental class can be used for this purpose, and this leads to the $I_{0,n,\beta}^X$ above. In other cases the situation is considerably more involved; the construction of Behrend and Fantechi produces a good candidate in great generality.

REMARK. Our previous Gromov-Witten number is still $\int_{\overline{M}_{0,n}} I_{0,n,\beta}^X(\gamma_1 \otimes \cdots \otimes \gamma_n)$.

Summarizing: the Gromov-Witten information is encoded in maps

$$I_{0,n,\beta}^X : (H^*X)^{\otimes n} \to H^*(\overline{M}_{0,n})$$

We can also put all of them together if we want, by taking as many copies of the target as there are effective β's:

$$I_{0,n}^X : (H^*X)^{\otimes n} \to \oplus_{\beta \in \mathcal{B}} H^*(\overline{M}_{0,n})_{(\beta)}$$

where $\mathcal{B} = \{\text{effective } \beta\text{'s}\}$, and the product on the $\oplus$ respects the $\mathcal{B}$-grading. Put it otherwise, we should consider

$$I_{0,n}^X : (H^*X)^{\otimes n} \to H^*(\overline{M}_{0,n})$$

as a map between $\mathcal{B}$-graded objects in order to carry along the information about β. *We will do this implicitly in what follows.*

§6. **GW on X induce $H_*\mathcal{M}(n) \to \mathbf{End}_{H^*X}(n)$.** At this point we have maps

$$(H^*X)^{\otimes n} \otimes H^*X \to H^*(\overline{M}_{0,n+1})$$

$$(\gamma_1 \otimes \cdots \otimes \gamma_n) \otimes \gamma_{n+1} \mapsto I(\gamma_1 \otimes \cdots \otimes \gamma_{n+1})$$

where $I(\gamma_1 \otimes \cdots \otimes \gamma_{n+1}) = \oplus_\beta I^X_{0,n+1,\beta}(\gamma_1 \otimes \cdots \otimes \gamma_{n+1})$ (again, source and target of I are $\mathcal{B}$-graded, see §5; this will be hidden in the notations). Dualizing this map, we obtain a map

$$H_*(\overline{M}_{0,n+1}) \to \mathrm{Hom}((H^*X)^{\otimes n} \otimes H^*X, \mathbb{Q})$$
$$= \mathrm{Hom}((H^*X)^{\otimes n}, (H^*X)^\vee)$$
$$= \mathrm{Hom}((H^*X)^{\otimes n}, H^*X)$$

by Poincaré duality. That is, we now have a map

$$H_*\mathcal{M}(n) \to \mathbf{End}_{H^*X}(n)$$

for $n \geq 2$, which we proceed to write out explicitly.

First, Poincaré duality works $H^*X \to (H^*X)^\vee$ by sending c to $\int_X c \cup \cdot = \alpha_c(\cdot)$. Say T_i form a basis for H^*X, and $(g_{ij}) = \int T_i \cup T_j$, $(g^{ij}) = (g_{ij})^{-1}$ as usual. Then, writing c in terms of this basis:

$$c = \sum c^i T_i \mapsto \left(T_j \mapsto \alpha_c(T_j) = \int_X \sum_i c^i T_i \cup T_j = \sum_i g_{ij} c^i \right) \quad , \text{ and hence}$$

$$c^k = \sum_i \delta_i^k c^i = \sum_{ij} g^{jk} g_{ij} c^i = \sum_j g^{jk} \alpha_c(T_j) \quad , \text{ or}$$

$$c = \sum_k c^k T_k = \sum_{jk} g^{jk} \alpha_c(T_j)\, T_k$$

Now start with $Z \in H_*\mathcal{M}(n) = H_*\overline{M}_{0,n+1} = (H^*\overline{M}_{0,n+1})^\vee$, and get

$$Z \mapsto \int_Z \cdot \mapsto \left(\gamma_1 \otimes \cdots \otimes \gamma_{n+1} \mapsto \int_Z I(\gamma_1 \otimes \cdots \otimes \gamma_{n+1}) \right) \in \mathrm{Hom}((H^*X)^{\otimes n+1}, \mathbb{Q})$$

$$= \left(\gamma_1 \otimes \cdots \otimes \gamma_n \mapsto \int_Z I(\gamma_1 \otimes \cdots \otimes \gamma_n \otimes \cdot) \right) \in \mathrm{Hom}((H^*X)^{\otimes n}, (H^*X)^\vee)$$

$$= \left(\gamma_1 \otimes \cdots \otimes \gamma_n \mapsto \sum_{jk} g^{jk} \int_Z I(\gamma_1 \otimes \cdots \otimes \gamma_n \otimes T_j)\, T_k \right)$$

in $\mathrm{Hom}((H^*X)^{\otimes n}, H^*X) = \mathbf{End}_{H^*X}(n)$. This defines

$$H_*\mathcal{M}(n) \to \mathbf{End}_{H^*X}(n)$$

for all $n \geq 2$. For $n = 1$, there is little choice:

$$H_*\mathcal{M}(1) = H_*(\mathrm{pt}) = \mathbb{Q} \to \mathrm{Hom}(H^*X, H^*X) = \mathbf{End}_{H^*X}(1)$$
$$1 \mapsto \text{identity}$$

§7. Properties of Gromov-Witten invariants and morphisms of operads. The (vague) claim is now that *the general properties of Gromov-Witten invariants amount to the fact that the maps*

(*) $$H_*\mathcal{M}(n) \to End_{H^* X}(n)$$

defined in §6 preserve the operad structures on source and target.

REMARK/EXAMPLE. Both operads involved have in fact a little more structure, and this is also preserved. On the End side, we have

$$\psi : End_{H^* X}(n) \to End_{H^* X}(n-1)$$
$$\alpha \mapsto (\gamma_1 \otimes \cdots \otimes \gamma_{n-1} \mapsto \alpha(\gamma_1 \otimes \cdots \otimes \gamma_{n-1} \otimes 1))$$

On the $H_*\mathcal{M}$ side, we have the maps induced in homology by the maps

$$\phi : \mathcal{M}(n) = \overline{M}_{0,n+1} \to \overline{M}_{0,n} = \mathcal{M}(n-1)$$

obtained by forgetting the last point (and collapsing unstable components):

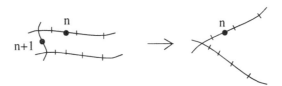

CLAIM. *The map $H_*\mathcal{M} \to End_{H^* X}$ defined above preserves this structure; that is, the diagrams*

$$
\begin{array}{ccc}
H_*\mathcal{M}(n) & \longrightarrow & End_{H^* X}(n) \\
\phi \downarrow & & \downarrow \psi \\
H_*\mathcal{M}(n-1) & \longrightarrow & End_{H^* X}(n-1)
\end{array}
$$

commute.

PROOF. These (and all other analogous facts) should be straightforward from the explicit formula given for (*) in §6. Going first right and then down gives

$$Z \mapsto \left(\gamma_1 \otimes \cdots \otimes \gamma_n \mapsto \sum_{jk} g^{jk} \int_Z I(\gamma_1 \otimes \cdots \otimes \gamma_n \otimes T_j) T_k \right)$$

$$\mapsto \left(\gamma_1 \otimes \cdots \otimes \gamma_{n-1} \mapsto \sum_{jk} g^{jk} \int_Z I(\gamma_1 \otimes \cdots \otimes \gamma_{n-1} \otimes 1 \otimes T_j) T_k \right) \quad ;$$

going first down and then right gives

$$Z \mapsto \phi_* Z \mapsto \left(\gamma_1 \otimes \cdots \otimes \gamma_{n-1} \mapsto \sum_{jk} g^{jk} \int_{\phi_* Z} I(\gamma_1 \otimes \cdots \otimes \gamma_{n-1} \otimes T_j) T_k \right)$$

Using that (g^{jk}) is nonsingular, reading components, and replacing T_j by γ_n we see then that the commutativity of the diagrams is equivalent to

$$\int_Z I(\gamma_1 \otimes \cdots \otimes \gamma_n \otimes 1) = \int_{\phi_* Z} I(\gamma_1 \otimes \cdots \otimes \gamma_n)$$

for all $Z \in H_* \overline{M}_{0,n+1}$ and $\gamma_1, \ldots, \gamma_n \in H^* X$. This holds by the projection formula, and is essentially equivalent to property (1) of Gromov-Witten invariants in Ranestad's lecture. $\square$

Concerning associativity, writing that (*) induces a morphism of operads also amounts to the commutativity of certain diagrams. Writing this out leads essentially to the WDVV equations, which also appeared in Ranestad's lecture. A key formula from that lecture, rewritten slightly to match notations, reads

$$\sum \int_{\overline{M}_{0,A \cup \{\bullet\}}} I((\otimes_{a \in A} \gamma_a) \otimes T_\ell) \, g^{\ell m} \int_{\overline{M}_{0,B \cup \{\bullet\}}} I(T_m \otimes (\otimes_{b \in B} \gamma_b)) = \int_{D(A|B)} I(\otimes_{c \in A \cup B} \gamma_c)$$

(for the notations, see also Belorousski's lecture on $\overline{M}_{0,n}$. And note that the dependence on β is implicit in I once the relevant objects are B-graded, see §5; the β-component of the left-hand-side is a $\sum_{\beta_1 + \beta_2 = \beta}$ of terms $\int I_{\beta_1} \cdots \int I_{\beta_2} \cdots$). We want to show that this follows from the commutativity of one of the diagrams expressing that (*) is a morphism of operads. Conversely, these equalities and the structure of the boundary of the spaces $\overline{M}_{0,n}$ ought to imply that *all* such diagrams commute; but this seems substantially more involved, and we will not attempt to discuss it here.

For $|A| = n_1$, $|B| = n_2$, with $n_1, n_2 \geq 2$, consider

$$\mathcal{M}(n_2) \times \mathcal{M}(n_1) \times \underbrace{\mathcal{M}(1) \times \cdots \times \mathcal{M}(1)}_{n_2 - 1} \to \mathcal{M}(n_1 + n_2 - 1)$$

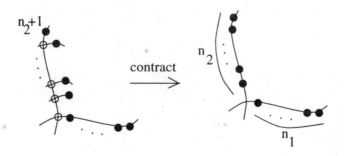

This map realizes the map

$$\overline{M}_{0,A\cup\{\bullet\}} \times \overline{M}_{0,B\cup\{\bullet\}} \to \overline{M}_{0,A\cup B}$$

whose image is the divisor $D(A|B)$ of $\overline{M}_{0,A\cup B}$. Now take H_* and apply $(*)$:

$$H_*\mathcal{M}(n_2) \otimes H_*\mathcal{M}(n_1) \otimes H_*\mathcal{M}(1)^{\otimes(n_2-1)} \longrightarrow H_*\mathcal{M}(n_1 + n_2 - 1)$$

$$\downarrow \qquad\qquad\qquad\qquad\qquad\qquad\qquad\qquad \downarrow$$

$$\mathrm{End}_{H^*X}(n_2)\otimes\mathrm{End}_{H^*X}(n_1)\otimes\mathrm{End}_{H^*X}(1)^{\otimes(n_2-1)} \longrightarrow \mathrm{End}_{H^*X}(n_1+n_2-1)$$

Assume $(*)$ induces a morphism of operads; then this diagram commutes. Chase

$$[\overline{M}_{0,B\cup\{\bullet\}}] \otimes [\overline{M}_{0,A\cup\{\bullet\}}] \otimes [\mathrm{pt}]^{\otimes n_2-1} \quad :$$

going first right and then down, this maps to $[D(A|B)]$ and then to

$$\gamma_1 \otimes \cdots \otimes \gamma_{n_1+n_2-1} \mapsto \sum_{jk} g^{jk} \int_{D(A|B)} I(\gamma_1 \otimes \cdots \otimes \gamma_{n_1+n_2-1} \otimes T_j)\, T_k \quad ;$$

going first down and then right, it maps to

$$\left(\sum_{jk} g^{jk} \int_{\overline{M}_{B\cup\{\bullet\}}} I(\underbrace{\cdot \otimes \cdots \otimes \cdot}_{n_2} \otimes T_j)\, T_k \right)$$

$$\otimes \left(\gamma_1 \otimes \cdots \otimes \gamma_{n_1} \mapsto \sum_{\ell m} g^{\ell m} \int_{\overline{M}_{A\cup\{\bullet\}}} I(\gamma_1 \otimes \cdots \otimes \gamma_{n_1} \otimes T_\ell)\, T_m \right)$$

$$\otimes (\gamma_{n_1+1} \mapsto \gamma_{n_1+1}) \otimes \cdots \otimes (\gamma_{n_1+n_2-1} \mapsto \gamma_{n_1+n_2-1})$$

and then (by the obvious linearity of I) to

$$\gamma_1 \otimes \cdots \otimes \gamma_{n_1+n_2-1} \mapsto \sum_{jk} g^{jk} \left(\sum_{\ell m} g^{\ell m} \int_{\overline{M}_{A\cup\{\bullet\}}} I(\gamma_1 \otimes \cdots \otimes \gamma_{n_1} \otimes T_\ell) \right.$$

$$\left. \cdot \int_{\overline{M}_{B\cup\{\bullet\}}} I(T_m \otimes \gamma_{n_1+1} \otimes \cdots \otimes \gamma_{n_1+n_2-1} \otimes T_j)\, T_k \right)$$

Comparing the two results, we see that we must have

$$\sum \int_{\overline{M}_{A\cup\{\bullet\}}} I(\gamma_1 \otimes \cdots \otimes \gamma_{n_1} \otimes T_\ell) g^{\ell m} \int_{\overline{M}_{B\cup\{\bullet\}}} I(T_m \otimes \gamma_{n_1+1} \otimes \cdots \otimes \gamma_{n_1+n_2-1} \otimes T_j)$$

$$= \int_{D(A|B)} I(\gamma_1 \otimes \cdots \otimes \gamma_{n_1+n_2-1} \otimes T_j)$$

for all $\gamma_1, \ldots, \gamma_{n_1+n_2-1}$, and therefore (again by linearity of I)

$$\sum \int_{\overline{M}_{A\cup\{\bullet\}}} I(\gamma_1 \otimes \cdots \otimes \gamma_{n_1} \otimes T_\ell) \, g^{\ell m} \int_{\overline{M}_{B\cup\{\bullet\}}} I(T_m \otimes \gamma_{n_1+1} \otimes \cdots \otimes \gamma_{n_1+n_2})$$

$$= \int_{D(A|B)} I(\gamma_1 \otimes \cdots \otimes \gamma_{n_1+n_2})$$

which is our basic identity. $\square$

7. Axioms for Gromov-Witten invariants, I—B. Fantechi
October 24, 1996

We introduce and motivate the axioms from [K-M] for Gromov-Witten invariants, then discuss approaches for a construction of classes satisfying these axioms, with emphasis on Behrend's work [B-M]. This talk and the next one are a rough and oversimplified outline of the contents of [K-M] and [B-M]. They contain *intentional mistakes* (and probably unintentional ones as well).

Let X be a smooth complex projective variety.

DEFINITION. A *system of Gromov-Witten classes* on X is the datum, for every $\beta \in H_2(X)$ and for every $g, n \geq 0$ such that $2g + n \geq 3$, of a linear map .

$$I_{g,n,\beta}^X : H^*(X)^{\otimes n} \to H^*(\overline{M}_{g,n})$$

satisfying the properties

(GW0) Effectivity
(GW1) S_n-equivariance
(GW2) Grading
(GW3) Fundamental class
(GW4) Divisors
(GW5) Mapping to a point ($\beta = 0$)
(GW6) Splitting
(GW7) Genus reduction
(GW8) Motivic axiom

A *tree level system* of Gromov-Witten classes is the same, with $g = 0$ (so the genus reduction axiom becomes irrelevant).

Here $H_2(X) = H_2(X, \mathbb{Z})$, and cohomology is taken with $\mathbb{Q}$-coefficients.

Idea behind the axioms: suppose we are in the best possible world; in particular, nontrivial finite groups of automorphisms do not exist, and for a generic map $f : C \to X$, $h^1(f^*T_X) = 0$. (Note: this is *never* true if $g(C) \geq 1$.) Let $\overline{M}_{g,n}(X, \beta)$ be the moduli space of maps (as defined in previous lectures). What is its dimension supposed to be? 'Coarse' reasoning: look at the map $\overline{M}_{g,n}(X, \beta) \to \overline{M}_{g,n}$, and pretend it simply forgets the extra data. The dimension of $\overline{M}_{g,n}$ is $(3g - 3 + n)$; we are pretending that there is no (or no generic) obstruction from H^1, so (by Riemann-Roch) the fibers would have dimension

$$h^0(g^*T_X) = \chi(f^*T_X) = \dim X(1 - g) + \beta \cdot c_1(X) \quad , \quad \text{hence}$$

$$\dim \overline{M}_{g,n}(X, \beta) = (\dim X - 3)(1 - g) + n + \beta c_1(X)$$

We call this number the *expected dimension* of $\overline{M}_{g,n}(X, \beta)$. The natural map

$$\overline{M}_{g,n}(X, \beta) \to X^n \times \overline{M}_{g,n} \quad :$$

gives (via Poincaré duality) a class

$$c_{g,n,\beta}^X = [\overline{M}_{g,n}(X, \beta)] \in H^*(X^n \times \overline{M}_{g,n})$$

Consider the two projection

$$X^n \times \overline{M}_{g,n} \xrightarrow{\quad q \quad} \overline{M}_{g,n}$$

$$p \downarrow$$

$$X^n$$

and define (note: $H^*(X^n) = H^*(X)^{\otimes n}$)

$$I^X_{g,n,\beta} = q_*(p^*(\cdot) \cap c^X_{g,n,\beta})$$

(motivated by the intuitive enumerative meaning).

We will now state each of the axioms and 'verify' some of them in the very optimistic assumptions above, in the hope to give an intuitive motivation for them.

(GW0) Let $H_2(X)_+ = \{\beta \in H_2(X) / \beta \cdot c_1(\mathcal{L}) \geq 0 \text{ for all ample } \mathcal{L}\}$. Then $I^X_{g,n,\beta} = 0$ unless $\beta \in H_2(X)_+$. Verification: $\overline{M}_{g,n}(X,\beta)$ is clearly empty unless either $\beta = 0$, or β is the class of an effective case. In either case, $\beta \in H_2(X)_+$.

(GW1) $I^X_{g,n,\beta}$ is S_n-equivariant.

(GW2) $\deg I^X_{g,n,\beta} = 2[(g-1)\dim X + \beta \cdot K_X]$. Verification: This is the degree induced by the dimension estimate above (the '2' comes in passing from the complex to the real dimension).

(GW3) Let $\pi : \overline{M}_{g,n+1} \to \overline{M}_{g,n}$ be the map forgetting the last point (and stabilizing); then $I^X_{g,n+1,\beta}(\cdot \otimes 1_X) = \pi^* \circ I^X_{g,n,\beta}$. Moreover,

$$I^X_{0,3,\beta}(\gamma_1 \otimes \gamma_2 \otimes 1_X) = \begin{cases} \int_X \gamma_1 \cup \gamma_2 & \text{if } \beta = 0 \\ 0 & \text{otherwise} \end{cases}$$

(GW4) If $\gamma \in H^2(X)$, then

$$\pi_* I^X_{g,n+1,\beta}(\cdot \otimes \gamma) = (\beta \cdot \gamma) I^X_{g,n,\beta}$$

(GW5) (Case $\beta = 0$) $\overline{M}_{g,n}(X,0) = \overline{M}_{g,n} \times X$. Here

$$\dim \overline{M}_{g,n}(X,0) = 3g - 3 + n + \dim X$$
$$\text{exp. } \dim \overline{M}_{g,n}(X,0) = 3g - 3 + n + \dim X(1-g)$$

do not match for $g > 0$. Now the obstruction space is:

$$T^1_{(C,x,f)} = H^1(C, f^*T_X) = H^1(C, \mathcal{O}_C) \otimes T_{x,f(C)} \quad ;$$

that is, with

$$\begin{array}{ccc} \mathcal{C} & \xrightarrow{\ f\ } & X \\ {\scriptstyle g}\downarrow & & \\ \overline{M}_{g,n}(X,0) & & \end{array} ,$$

$T^1 = R^1 g_*(f^* T_X)$. This has rank $g \dim X$, accounting for the difference between actual and expected dimension. Use T^1 then to 'correct' the fundamental class:

$$c^X_{g,n,0} = [\overline{M}_{g,n}(X,0)] \cup c_{\text{top}} T^1 \quad .$$

Note: in the real world, $\mathcal{C}$ does not exist; this is one reason to consider $\overline{M}_{g,n}(X,\beta)$ as a Deligne-Mumford stack: so it is a smooth, fine moduli space, can work with cohomology (with $\mathbb{Q}$-coefficients), etc. Define

$$I^X_{g,n,0} = p_{2*}(p_1^*(\cdot) \cdot c^X_{g,n,0})$$

where p_i are the projections from $\overline{M}_{g,n} \times X$.

(GW6). Fix g_1, g_2, g, n_1, n_2, n such that $2g_i + n_i + 1 \geq 3$, $g = g_1 + g_2$, $n = n_1 + n_2$. Let $\varphi : \overline{M}_{g_1,n_1+1} \times \overline{M}_{g_2,n_2+1} \to \overline{M}_{g,n}$ be the usual glueing map. Then

$$\varphi^* \circ I^X_{g,n,\beta}(\gamma_1 \otimes \cdots \otimes \gamma_n) = \sum_{\beta_1+\beta_2=\beta} I^X_{g_1,n_1+1,\beta_1} \otimes I^X_{g_2,n_2+1,\beta_2}(\gamma_1 \otimes \cdots \otimes \gamma_{n_1} \otimes [\Delta] \otimes \gamma_{n_1+1} \otimes \cdots \otimes \gamma_n)$$

where $[\Delta]$ is the class of the diagonal in X^2 (this can also be written in terms of a basis and of the usual g^{ab}). Note that the sum is finite, by effectivity.

(GW7) Let $\psi : \overline{M}_{g-1,n+2} \to \overline{M}_{g,n}$ be the map joining two of the points and increasing the genus. Then

$$\psi^* \circ I^X_{g,n,\beta} = I^X_{g-1,n+2,\beta}(\cdot \otimes [\Delta])$$

(GW8) There exists $c^X_{g,n,\beta} \in A_*(X^n \times \overline{M}_{g,n})$ such that

$$I^X_{g,n,\beta} = q_*(p^*(\cdot) \cap c^X_{g,n,\beta})$$

with q, p the projections, as before.

REMARK. Once we take (GW8) for granted, all other axioms can be formulated in terms of $c^X_{g,n,\beta}$. For example, both (GW3) and (GW4) boil down to

$$c^X_{g,n+1,\beta} = (id_{X^n} \times \pi)^* c^X_{g,n,\beta}$$

We now try to describe how to construct a system of classes $c^X_{g,n,\beta}$ such that the axioms hold.

Constructions. If $g = 0$ and X is convex, then everything works fine, with $c^X_{0,n,\beta} = [\overline{M}_{0,n}(X, \beta)]$. The expectation is that this should be the case whenever the dimension equals the expected dimension. *Main problem: define $[\overline{M}_{g,n}(X, \beta)]^{\mathrm{virt}} \in A_*(\overline{M}_{g,n}(X, \beta))$ with the correct dimension and properties.*

Symplectic approach. Deform enough data so that $\overline{M}_{g,n}(X, \beta)$ is of the correct dimension. This requires a sophisticated analysis. Solutions have been given by Fukaya-Ono [F-O], Li-Tian [Li-Tian2], Siebert [Siebert].

Algebraic approach, Li & Tian's method [Li-Tian]. We first describe the situation locally in the euclidean topology. Let M be a moduli space. Usually from deformation theory one has for all $m \in M$ an obstruction space $O_{M,m}$. Choose m_0, and let $T = T_{M,m_0}$, $O = O_{M,m_0}$. Then there is a map of germs of analytic spaces $f : (T, 0) \to (O, 0)$ such that $(M, m_0) = f^{-1}(0)$; for $m \in (M, m_0)$, $T_{M,m} = \ker df(m)$ and $O_{M,m} = \mathrm{Coker}\, df(m)$. Then Li and Tian define a normal cone

$$C_{M/T} \subset O \times M \quad ,$$

of pure dimension $= \dim T$, and T-invariant fiberwise.

Global story: Given $E_T \overset{g}{\to} E_O$ morphism of vector bundles over M, such that $T_{M,m} = \ker g(m)$, $O_{M,m} = \mathrm{Coker}\, g(m)$ (the obstruction complex), there is a cone $C \subset E_0$ of pure dimension $= \mathrm{rk}\, T$, and we can define

$$[M]^{\mathrm{virt}} = [C] \cup (\text{zero section of } E_0)$$

Note $\dim[M]^{\mathrm{virt}} = \mathrm{rk}\, E - \mathrm{rk}\, F = \dim T(m) - \dim O(m)$ (=expected dimension) for all m.

The hard work goes now in defining everything rigorously and proving the independence on the choices, and the relevant properties.

Behrend-Fantechi: For all m, C_m comes from a cone in $O(m)$.

(1) The cone only depends only on the deformation functor;

(2) invariance means it comes from $[E_O/E_T]$ (stack quotient);

(3) The stack is an easy version of the deformation functor.

Result: given any M, Deligne-Mumford stack, there exists a pure dimensional $\underline{C}_M$ (the *intrinsic normal cone*, see [B-F]); and given any $E_T \to E_O$, the obstruction complex $\underline{C}_M \hookrightarrow [E_O/E_T] = \underline{E}$ yields a $[M]^{\mathrm{virt}} = [\underline{C}_M] \cap$ zero-section of $\underline{E}$. This only depends on $[E_O/E_T]$, i.e., on $E_T \to E_O$ as an object in the derived category.

There is also a relative version over any smooth Artin stack.

DEFINITION. For any g, any n, any X and any β, let $\mathcal{M} = \mathcal{M}_{g,n}$, the moduli stack of prestable curves of genus g with n marked points: $\mathcal{M}$ is a smooth Artin stack, containing $\overline{M}_{g,n}$ as an open and proper subset ($\mathcal{M}$ is not separated). There is a natural morphism $\overline{M} := \overline{M}_{g,n}(X, \beta) \to \mathcal{M}$, sending (C, x_i, f) to (C, x_i) (without stabilizing). Let $\mathcal{C} \to \overline{M}$ be the universal curve, and consider the diagram

$$C \xrightarrow{f} X$$
$$p \downarrow$$
$$\overline{M}$$

The complex $Rp_*(f^*T_X)$, a well-defined object in the derived category, is a relative obstruction theory for $\overline{M}$ over $\mathcal{M}$, and yields a $[\overline{M}_{g,n}(X,\beta)]^{\text{virt}}$ of the correct dimension.

This implies immediately (GW0), (GW1), (GW2), (GW8); and (GW5) with a little care. To prove (GW3) and (GW4), one must consider

$$\overline{M}_{g,n+1}(X,\beta) \xrightarrow{\pi} \overline{M}_{g,n}(X,\beta)$$

The map π is flat of relative dimension 1. Then one needs

$$[\overline{M}_{g,n+1}(X,\beta)]^{\text{virt}} = \pi^*[\overline{M}_{g,n}(X,\beta)]^{\text{virt}}$$

This follows from properties of the relative intrinsic normal cone.

To prove (GW6) and (GW7), and for any further work we will have to face *graphs* and axioms from [B-M].

---

8. Axioms for Gromov-Witten invariants, II—B. Fantechi
October 31, 1996

This lecture is based on [B-M] and [Behrend] (the same disclaimer for the previous lecture applies here). Among the problems left open from the first lecture, we had the *splitting axiom,* dealing with the behavior of the Gromov-Witten classes on sets of reducible curves. This leads to considering more complicated curves.

DEFINITION. A *modular graph* τ is the datum of

(1) a finite set V of *vertices;*
(2) a finite set F of *flags;*
(3) a map $\partial : F \to V$;
(4) an involution $j : F \to F$ (denote $\overline{f} = j(f)$)
(5) (modular) a map, the *genus,* $g : V \to \mathbb{Z}_{\geq 0}$

Notation: $T = $ set of *tails* $= \{f \in F/\overline{f} = f\}$; $E = $ set of *edges* $= \{\{f, \overline{f}\}/f \neq \overline{f}\}$; for $v \in V$, $F_v = \{f/\partial f = v\}$.

DEFINITION. The *topological realization* of τ, denoted $|\tau|$: start with a point for every vertex; for every edge $\{e, \overline{e}\}$ glue $[0,1]$ at ∂e, $\partial \overline{e}$; for every tail t, glue $[0,1]$ at ∂t.

Intuitive relation with pointed curves: every vertex v corresponds to a curve of genus $g(v)$; for every edge we make the corresponding curves intersect transversally; for every tail we put a marked point. Hence a modular graph describes the structure of a prestable curve.

DEFINITION. A vertex v of τ is *stable* if $2g(v) + \#F_v \geq 3$.

We will assume all vertices of all graphs under consideration to be stable.

EXAMPLE. An irreducible n-pointed genus-g curve corresponds to a single vertex v, with $g(v) = g$, and $F = F_v = \{1, \ldots, n\}$.

Fix a smooth projective variety X over $\mathbb{C}$.

DEFINITION. Let τ be a modular graph. An X-marking of τ is a map $\alpha : V(\tau) \to H_2(X)^+$.

DEFINITION. Fix (τ, α). A *prestable* (τ, α)-*map to* X is the datum (C_v, x_f, μ), where

(1) $\forall v \in V(\tau)$, C_v is a prestable curve of genus $g(v)$;
(2) $\forall f \in F(\tau)$, $x_f \in C_{\partial f}$;
(3) μ maps $C = \amalg C_v \to X$ so that $\mu(x_f) = \mu(x_{\bar{f}})$ for all $f \in F$; and
$\mu_*[C_v] = \alpha(v) \in H_2(X)^+$.

Moreover, (C_v, x_f, μ) is *stable* if $\forall v \in V$, $(C_v, \{x_f\}_{f \in F_v}, \mu_{C_v})$ is a stable map.

REMARK. If $X =$ pt, α and μ bear no information. We speak then simply of a τ-(pre)stable curve.

REMARK. Note that the domain of a prestable (τ, α) map has at least as many irreducible components as the vertices of τ, but it might have more (i.e., some curve C_v may be reducible, or singular).

DEFINITION. A family of (τ, α)-prestable maps over S is the datum (C_v, x_f, μ) where

(1) $C_v \to S$ is a flat proper map, and μ is a map $C = \amalg C_v \to X$;
(2) $x_f : S \to C_v$ is a section (for $f \in F_v$), and $\mu \circ x_f = \mu \circ x_{\bar{f}}$;
(3) $\forall s \in S$, the fiber of (C_v, x_f, μ) over s is a (τ, α)-prestable curve.

REMARK. Families of (τ, α) (pre)stable maps pull-back; so we may consider the corresponding functor, as usual, and as usual we wish to represent it.

THEOREM. *Let X be a projective variety, (τ, α) an X-marked modular graph.*

(1) *There exists a fine moduli space $\overline{M}_\tau(X, \alpha)$ for (τ, α)-stable maps to X; $\overline{M}_\tau(X, \alpha)$ is a proper, separated Deligne-Mumford stack of finite type.*
(2) *There exists a fine moduli space $\mathcal{M}(\tau)$ for τ-prestable curves; $\mathcal{M}(\tau)$ is a smooth Artin stack.*

For example, for τ a graph with a single vertex of genus g and n flags around it, $\overline{M}_\tau(X, \alpha)$ is our usual $\overline{M}_{g,n}(X, \alpha)$.

DEFINITION. Let (τ, α) be a marked graph.

(1) $\chi(\tau) = \chi(|\tau|) - \sum_{v \in V(\tau)} g(v)$.
(2) $g(\tau) = 1 - \chi(\tau)$.

(3) $\dim \tau = -3\chi(\tau) + \#T - \#E$.
(4) The class of $\tau = \alpha(\tau) = \sum_{v \in V} \alpha(v)$.
(5) $\dim(\tau, \alpha) = \dim(\tau) + \chi(\tau) \dim X + \alpha(\tau) c_1(T_X)$.

Exercise: prove that $\overline{M}_\tau$ has dimension $= \dim(\tau)$.
There are natural maps

$$\overline{M}_\tau(X, \alpha) \to \mathcal{M}(\tau) \to \overline{M}_\tau$$

given by $(C_v, x_f, \mu) \mapsto (C_v, x_f) \mapsto (C_v, x_f)^{\text{stab}}$.
    The *expected dimension* of $\overline{M}_\tau(X, \alpha)$ is $\dim(\tau, \alpha)$.

DEFINITION. An *orientation* for $\overline{M}$ over X is the datum, for every X-marked graph (τ, α), of a class

$$\mathcal{I}_\tau(X, \alpha) \in A_{\dim(\tau, \alpha)} \overline{M}_\tau(X, \alpha)$$

satisfying 5 compatibility axioms, to be discussed shortly.

    An orientation determines a good system of Gromov-Witten classes, in the sense of the previous lecture, as follows. Omitting X from the notations for convenience, we have the diagram:

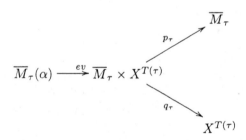

and we set $\tilde{\mathcal{I}}_\tau(\alpha) = ev_* \mathcal{I}_\tau(\alpha)$, and define

$$I_\tau^\alpha : H^* \otimes T(\tau) \to H^*(\overline{M}_\tau) \quad \text{by}$$

$$I_\tau^\alpha(\gamma) = p_{\tau *}\left(q_\tau^*(\gamma) \cap \tilde{\mathcal{I}}_\tau(\alpha) \right) \quad .$$

    We list the compatibility axioms and give for each of them a short, informal description.
(BM1) Mapping to a point
(BM2) Products
(BM3) Cutting edges
(BM4) Forgetting tails
(BM5) Isogeny

(BM1) This is the case $\alpha = 0$ ($\Longleftrightarrow$ $\alpha(\tau) = 0$). The axiom is the obvious generalization of the corresponding GW axiom (GW5).

(BM2) If (τ, α) is the 'disjoint union' of (τ_1, α_1) and (τ_2, α_2), then $\mathcal{I}_\tau(\alpha) = \mathcal{I}_{\tau_1}(\alpha_1) \times \mathcal{I}_{\tau_2}(\alpha_2)$.

(BM3) Cutting an edge: the modular graph σ is obtained from τ by 'cutting the edge' $\{e, \bar{e}\}$ if $V(\sigma) = V(\tau)$, $F(\sigma) = F(\tau)$, $\partial\sigma = \partial\tau$, $g_\sigma = g_\tau$, and

$$j_\sigma(f) = \begin{cases} j_\tau(f) & f \notin \{e, \bar{e}\} \\ f & f \in \{e, \bar{e}\} \end{cases}$$

(so the edge $\{e, \bar{e}\}$ between internal vertices is replaced by two tails). We have the cartesian diagram

$$\begin{array}{ccc} \overline{M}_\tau(\alpha) & \longrightarrow & \overline{M}_\sigma(\alpha) \\ \downarrow & & \downarrow \\ X & \stackrel{\Delta}{\longrightarrow} & X^2 \end{array}$$

where the first vertical map sends (C_v, x_f, μ) to $\mu(x, e) = \mu(x, \bar{e})$, while the second sends (C_v, x_f, μ) to the pair $(\mu(x, e), \mu(x, \bar{e}))$. The axiom states then that the orientation for τ comes from σ:

$$\mathcal{I}_\tau(\alpha) = \Delta^!(\mathcal{I}_\sigma(\alpha)) \quad .$$

This corresponds to gluing two curves at one marked point on each.

(BM4) Forgetting tails: σ is obtained from τ by forgetting the tail $t \in T(\tau)$ if $V(\sigma) = V(\tau)$, $F(\sigma) = F(\tau) - \{t\}$, $j_\sigma = j_\tau$, $g_\sigma = g_\tau$, $\partial_\sigma = \partial_\tau$ (warning: we are cheating a little because σ here could become unstable; pretend it stays stable, for simplicity). This situation generalizes the map from $\overline{M}_{g,n}$ forgetting one of the marked points: we get a flat, proper map

$$\overline{M}_\tau(X, \alpha) \stackrel{\pi}{\to} \overline{M}_\sigma(X, \alpha)$$

and the axiom states that

$$\mathcal{I}_\tau(\alpha) = \pi^* \mathcal{I}_\sigma(\alpha) \quad .$$

This corresponds to forgetting one of the marked points.

(BM5) Isogeny: for simplicity, we will consider here only the case of 'contracting an edge'. The modular graph σ is obtained from τ by contracting the edge $\{e, \bar{e}\}$ if $F(\sigma) = F(\tau) - \{e, \bar{e}\}$, $\partial_\sigma = \partial_\tau$, $j_\sigma = j_\tau$, and

—if $\{e, \bar{e}\}$ forms a loop in τ, that is $\partial e = \partial \bar{e}$, then $V(\sigma) = V(\tau)$ and $g_\sigma(\partial e) = g_\tau(\partial e) + 1$;

—if $\{e, \bar{e}\}$ does not form a loop in τ, $V(\sigma) = (V(\tau) - \{\partial e, \partial \bar{e}\}) \cup \{*\}$, with $g_\sigma(*) = g_\tau(\partial e) + g_\tau(\partial \bar{e})$

(and $g_\sigma = g_\tau$ on unaffected vertices).

This corresponds to the smoothing of a node.

Further, for α a marking on σ consider all possible compatible markings α_i on τ. We get a diagram

$$
\begin{array}{ccc}
\amalg_i \overline{M}_\tau(X,\alpha_i) & \longrightarrow & \overline{M}_\sigma(X,\alpha) \\
\downarrow & & \downarrow \\
\overline{M}_\tau & \xrightarrow{\ \phi\ } & \overline{M}_\sigma
\end{array}
$$

and the axiom (in a slightly simplified form) prescribes that

$$
\sum \mathcal{I}_\tau(\alpha_i) = \phi^! \mathcal{I}_\sigma(\alpha) \quad .
$$

As mentioned above, it can be shown that an orientation defines a system of Gromov-Witten classes satisfying the axioms (GW0)—(GW8) of the previous lecture. For example, the splitting axiom compares classes for $\overline{M}_{g_1,n_1+1}$, $\overline{M}_{g_2,n_2+1}$ with classes for $\overline{M}_{g,n}$ with $g = g_1 + g_2$ and $n = n_1 + n_2$:

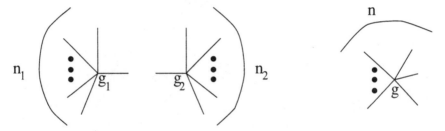

These are related to each other by respectively cutting an edge or contracting it from

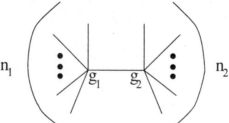

Thus the 'splitting axiom' (GW6) follows by judiciously applying (BM3) and (BM5). Similarly,

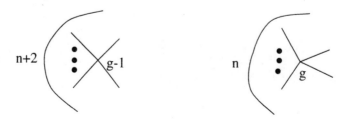

are obtained by respectively cutting the loop, or contracting it, in

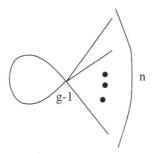

(BM3) and (BM5) can then be used to prove the 'genus reduction axiom' (GW7) for Gromov-Witten classes.

MAIN RESULT. *(Behrend) The class defined by*

$$\mathcal{I}_\tau(\alpha) = [\overline{M}_\tau(\alpha), E^\bullet_{\tau,\alpha}]^{\text{virt}}$$

gives an orientation for $\overline{M}$ over X.

In this statement, $[\cdot]^{\text{virt}}$ is the (relative) virtual fundamental class in the sense of the previous lecture, and $E^\bullet_{\tau,\alpha}$ is the relative obstruction theory over $\mathcal{M}_\tau$: $E^\bullet_{\tau,\alpha} = R\pi_*\mu^* T_X$ in

$$\begin{array}{ccc} \mathcal{C} & \xrightarrow{\ \mu\ } & X \\ \pi \downarrow & & \\ \overline{M}_\tau(X,\alpha) & & \end{array} \quad,$$

where $(\mathcal{C}, \mu, X)$ is the universal family over $\overline{M}_\tau(X,\alpha)$.

This result produces then a system of Gromov-Witten classes satisfying the axioms from [K-M], for an arbitrary smooth, projective variety X.

---

9. Kapranov's work on $\overline{M}_{0,n}$—C. Faber
October 29, 1996

The reference for this is [Kapranov]; we work over $\mathbb{C}$.

§1. The main result.

THEOREM. *Let $p_1, \ldots, p_n$ be n points in $\mathbb{P}^{n-2}$ in general position. Let $V_0(p) = V_0(p_1, \ldots, p_n)$ be the space of Veronese curves in $\mathbb{P}^{n-2}$ through $p_1, \ldots, p_n$. Then:*

(a) $V_0(p) \cong M_{0,n}$.

Consider $V_0(p)$ as a subvariety of the Hilbert scheme $\mathcal{H}$ parametrizing all subschemes of $\mathbb{P}^{n-2}$; let $V(p) = V(p_1, \ldots, p_n)$ be the closure of $V_0(p)$ in $\mathcal{H}$. Then:

(b) $V(p) \cong \overline{M}_{0,n}$. The subschemes of $\mathbb{P}^{n-2}$ representing points of $V(p)$ are, considered together with the n points, stable n-pointed curves of genus 0.

(c) Analogous statement for Chow variety instead of Hilbert scheme.

REMARKS. In suitable coordinates, $p_i = e_i = (0 : \cdots : 0 : \underset{i}{1} : 0 : \cdots : 0)$ for $0 \le i \le n-1$, and $p_n = e_n = (1 : \cdots : 1)$. Also, by a classical result of Castelnuovo, through any $(n+1)$ points in $\mathbb{P}^{n-2}$ in general position passes a unique Veronese curve.

§2. Discussion. Fix $p_1, \ldots, p_n$ in general position in $\mathbb{P}^{n-2}$, $\overline{M}_{0,n} \cong V(p)$.

CLAIM. *Every component of the curve $C \in V(p)$ is a Veronese curve in its span.*

A more precise statement requires a few preliminaries:

(a) Let $\mathcal{A}_{n-1}$ be the configuration of $\binom{n}{2}$ hyperplanes Λ_{ij}, $\Lambda_{ij} =$span of$\{p_k\}$, $k \ne i,j$ $(i \ne j)$. A *face* of $\mathcal{A}_{n-1}$ is a projective subspace which is the intersection of some of the Λ_{ij}.

The hyperplanes in $\mathcal{A}_{n-1}$ are isomorphic to the projectivization of the hyperplanes $\{t_i = t_j\}$ in the subspace $\mathbb{C}_0^{n-1} \subset \mathbb{C}^n$ with equation $t_1 + \cdots + t_n = 0$. (The 'mirror' of the root system A_{n-1}.)

(b) The faces in $\mathcal{A}_{n-1}$ correspond to equivalence relations on $\underline{n} = \{1, 2, \ldots, n\}$: for a relation R, $\Lambda(R) = \cap_{iRj} \Lambda_{ij}$.

(c) So there are exactly $2^{n-1} - 1$ 0-dimensional faces, corresponding to equivalence relations with exactly 2 equivalence classes (e.g., $p_i \leftrightarrow \{i\} \amalg \{\underline{n} - \{i\}\}$).

(d) Intersecting Λ_{ij} with a face Λ, we obtain an $\mathcal{A}_m$, with $m = \dim \Lambda + 1$.

(e) Let $\mathcal{T}$ be a tree with tails (=endpoints) $A_1, \ldots, A_n$, and let v, w be vertices (possibly endpoints) of $\mathcal{T}$. Let $[v, w]$ be the unique geodesic from v to w. Define for v an internal vertex of $\mathcal{T}$ an equivalence relation $\cong_v$ on $\underline{n}$:

$$i \cong_v j \iff v \notin [A_i, A_j]$$

The equivalence classes of $\cong_v$ correspond to the edges incident at v.

(f) For e an edge in $\mathcal{T}$, define

$$i \cong_e j \iff e \notin [A_i, A_j]$$

Note: for each e, this equivalence relation has exactly two equivalence classes.

(g) With the above understood, recall that every n-pointed stable curve corresponds to a tree (see Belorousski's lecture on $\overline{M}_{0,n}$). Then the claim is:

THEOREM. *For $C \in V(p)$, consider its tree $\mathcal{T}$. Let v be an internal vertex of $\mathcal{T}$, and let C_v be the corresponding component of C. Then*

(1) *C_v is a Veronese curve in its span $< C_v >$; $< C_v >$ is the face of $\mathcal{A}_{n-1}$ corresponding to $\cong_v$. Its dimension is*

(# equivalence classes)$-2=$(# edges incident to v)-2

$$=(\# \textit{ special points on } C_v)-2$$

(2) *Let e be an edge of T connecting internal vertices; that is, e corresponds to a node z of C. Then z corresponds to the 0-dimensional face of $\mathcal{A}_{n-1}$ corresponding to $\cong_e$. Therefore, the possible singular points of any $C \in V(p)$ belong to a fixed finite subset of $2^{n-1} - 1 - n$ points in $\mathbb{P}^{n-2}$.*

Next, recall that for $i = 1, \ldots, n$ there is a map $\pi_i : \overline{M}_{0,n} \to \overline{M}_{0,n-1}$ forgetting the i-th point. Thus there must be corresponding maps $V(p_1, \ldots, p_n) \to V(q_1, \ldots, q_{n-1})$.

CLAIM. *These are induced by the projections*

$$\mathbb{P}^{n-2} - \{p_i\} \dashrightarrow \mathbb{P}_i^{n-3}$$

This seems rather reasonable.

Next, there are basic line bundles $\mathcal{L}_i$ whose fiber at C is $T_{x_i}^* C$. These can be realized as follows: consider the map:

$$\sigma_i : V(p) \to \mathbb{P}_i^{n-3}$$

sending C to the embedded tangent line to C at p_i (which determines a point in $\mathbb{P}^{n-3}$ under the i-th projection as above). Then $\mathcal{L}_i = \sigma_i^*(\mathcal{O}_{\mathbb{P}_i^{n-3}}(1))$. Also, we get maps $\gamma_{\mathcal{L}_i} : \overline{M}_{0,n} \dashrightarrow \mathbb{P}H^0(\overline{M}_{0,n}, \mathcal{L}_i)^*$.

Results and comments:

(1) $\dim H^0(\overline{M}_{0,n}, \mathcal{L}_i) = n - 2$;
(2) $\gamma_{\mathcal{L}_i}$ is a birational morphism; $\gamma_{\mathcal{L}_i} = \sigma_i$ after the identification $\mathbb{P}_i^{n-3} = \mathbb{P}(H^0(\overline{M}_{0,n}, \mathcal{L}_i)^*$;
(3) σ_i can be decomposed explicitly into blow-ups; there are other constructions for this, due to Fulton-MacPherson and Keel;
(4) the identifications induce rational maps $\mathbb{P}_i^{n-3} \dashrightarrow \mathbb{P}_j^{n-3}$; these turn out to be Cremona transformations.
(5) Finally, this gives an interpretation for the Witten τ-numbers in genus 0:

$$< \tau_{d_1} \cdots \tau_{d_n} > = \int_{\overline{M}_{0,n}} \prod_i c_1(\mathcal{L}_i)^{d_i}$$

with $\sum d_i = n - 3$. This is the number of Veronese curves through a certain assortment of points, and tangent to certain codimension-d_i planes at these points. This might lead to a computation of these numbers (which are however already known and rather easy to obtain).

Next, we have the relative dualizing sheaf ω_C on a stable n-pointed curve C of genus 0: obtained by gluing ω_i on components C_i, where the ω_i are regular on the smooth part, and at a point of intersection of two components they may have simple poles with opposite residues. By Knudsen's work, $\omega_C(x_1 + \cdots + x_n)$ is very ample, and has $(n - 1)$ independent global sections. Via the corresponding map $C \hookrightarrow \mathbb{P}^{n-2}$, the images p_i of the x_i are in general position (as shown by a computation).

Now for part (a) of the main theorem, consider $x_1, \ldots, x_n$ distinct points on $\mathbb{P}^1$, and embed $\mathbb{P}^1$ in $\mathbb{P}^{n-2}$ with $\omega_{\mathbb{P}^1}(x_1 + \cdots + x_n)$. By a projective transformation, the image is moved to a Veronese curve through $p_1, \ldots, p_n$, given points in general position in $\mathbb{P}^{n-2}$. So each curve $\in M_{0,n}$ can be realized as stated. If f is an isomorphism of two n-pointed curves, f induces an isomorphism of the $(n-2)$-nd symmetric products $\check{\mathbb{P}}^{n-2}$, hence of their duals $\mathbb{P}^{n-2}$, fixing n generic points; it follows that f is the identity. This essentially establishes $M_{0,n} \cong V_0(p_1, \ldots, p_n)$.

For part (b), consider a stable n-pointed genus-0 curve over any base S: $C \xrightarrow{\pi} S$, with sections s_i. We have an embedding of C in $\mathbb{P}(\pi_*(\omega_{C/S}(s_1 + \cdots + s_n))^*)$, a projective bundle with n sections $s_1, \ldots, s_n$, in general position in every fiber. Trivialize the situation by moving every fiber to pass through a fixed frame in $\mathbb{P}^{n-2}$. The map π is flat, so we get a map $\gamma_{S,C}$ from S to the Hilbert scheme $\mathcal{H}$, such that C is the pull-back via $\gamma_{S,C}$ of the universal flat family of subschemes Ξ over $\mathcal{H}$:

$$
\begin{array}{ccc}
C & \longrightarrow & \Xi \\
\pi \downarrow & & \downarrow \\
S & \xrightarrow{\gamma_{S,C}} & \mathcal{H}
\end{array}
$$

Now for all S this determines a map

$$\gamma_S : \mathrm{Mor}(S, \overline{M}_{0,n}) \to \mathrm{Mor}(S, \mathcal{H})$$

by sending a family C (which determines and is determined by a unique morphism $S \to \overline{M}_{0,n}$) to $\gamma_{S,C}$. In fact, taking $S = \overline{M}_{0,n}$ itself gives a map from this to $\mathcal{H}$, and the dense open $M_{0,n}$ of $\overline{M}_{0,n}$ maps to $V_0(p) \subset \mathcal{H}$; so $\overline{M}_{0,n}$ maps to the closure $V(p)$ of $V_0(p)$ in $\mathcal{H}$, and γ_S must factor through

$$\gamma_S : \mathrm{Hom}(S, \overline{M}_{0,n}) \to \mathrm{Hom}(S, V(p))$$

We want to show that this is a bijection for all S. The injectivity follows from the fact that every C is a pull-back from Ξ; for the surjectivity, we have the diagram

$$
\begin{array}{ccc}
\overline{C}_{0,n} & \longrightarrow & \Xi \\
\pi \downarrow & & \downarrow \\
\overline{M}_{0,n} & \longrightarrow & V(p)
\end{array}
$$

and since the fibers of Ξ over V are stable n-pointed curves of genus 0, there must be a map $V(p) \to \overline{M}_{0,n}$ as $\overline{M}_{0,n}$ is a fine moduli space. $\square$

Note that we are assuming already that $\overline{M}_{0,n}$ is a fine moduli space; it would be interesting to use the construction to show directly that $V(p)$ is a fine moduli space.

For (c) in the main theorem, let C be the Chow variety; we have $\mathcal{H} \to C$ and $\overline{M}_{0,n} \cong V(p)$ surjecting on the closure $W(p)$ of $V_0(p)$ in C; we would show that this latter map is an isomorphism. One proves that it is a set bijection, then argues that tangent vectors to $V(p)$ (which is smooth) are not contracted in $W(p)$.

§3. **Problem.** Let $x_1(t), \ldots, x_n(t) \in \mathbb{C}((t))$ be distinct formal Laurent series in t. Assume that for $t \neq 0$, $(\mathbb{P}^1; x_1(t), \ldots, x_n(t))$ is a stable n-pointed curve in $M_{0,n} \subset \overline{M}_{0,n}$. The limit $\lim_{t \to 0}(\mathbb{P}^1; x_1(t), \ldots, x_n(t))$ is then a certain $C \in \overline{M}_{0,n}$; the question is: can this C be determined?

Choosing a suitable coordinate on $\mathbb{P}^1$, and multiplying by a common power of t if necessary, we may assume that $x_n(t)$ is identically ∞, and that $x_1(t), \ldots,$ $x_{n-1}(t) \in \mathbb{C}[[t]]$. Baby example: something like

$$(t^2, t^3, 2t^3, 3t^3, \infty)$$

Set-up. We define a tree, the 'tree of infinitely near points in $\mathbb{C}$': the vertices will be pairs (m, f) with $m \in \mathbb{Z}_{\geq -1}$, and f a polynomial in $\mathbb{C}[t]$ of degree $\leq m$ (the only polynomial of degree ≤ -1 is 0); the edges will be between pairs (m, f) and $(m+1, g)$ if and only if g agrees with f modulo t^{m+1}.

The ends of this tree correspond to formal Taylor series $\in \mathbb{C}[[t]]$.

For $(x_1(t), \ldots, x_{n-1}(t), \infty)$ as prescribed above, look at the subtree T of the tree T_0 of infinitely near points in $\mathbb{C}$ obtained as the union of paths in T_0 connecting the $x_i(t)$ and the root $v_0 = (-1, 0)$; and let T' be the simplest topologically equivalent version of T (where only internal vertices of degree ≥ 3 survive). Then:

CLAIM. *(1) T' is the tree of the limit curve;*

(2) to determine the isomorphism class of the limit, one needs the projective equivalence class of the special points on every component C_v of C. This can be read off from T': the edges departing from v correspond to the special points on C_v, and each has a number naturally associated with it (∞ for the edge connecting back towards the root, and the coefficients of the next term in the Taylor series represented by the other edges).

In the example above: $(t^2, t^3, 2t^3, 3t^3, \infty)$:

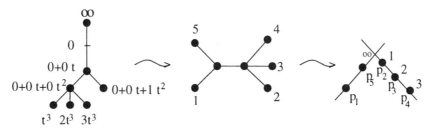

EXERCISE. For a slightly different flavor, take $K = \mathbb{Q}_2$, with ring of integers $\mathbb{Z}_2$, and determine the stable reduction (mod 2) of $(\mathbb{P}^1; 0, 1, 2, 3, 4, 5, 6, 7, 8)$.

10. Enumeration of rational curves, after Kontsevich—C. Faber
November 28, 1996

Three examples:

(1) Rational curves on $\mathbb{P}^2$. Consider the map $\varphi_k : \overline{M}_{0,k}(\mathbb{P}^2, d) \to (\mathbb{P}^2)^k$ defined by $\varphi_k(C; x_i; f) = (f(x_i))$, and let

$P_d := \#\{$rational curves on $\mathbb{P}^2$ of degree $d \geq 1$, through $(3d-1)$ generic points$\}$

=degree of φ_{3d-1}

$= \int_{\overline{M}_{0,3d-1}(\mathbb{P}^2,d)} \prod_{i=1}^{3d-1} \varphi^* \left(c_1(\mathcal{O}(1)_i)^2 \right)$

with evident notations.

(2) Rational curves on quintic threefolds. Let V be a threefold given by a section Q of $\mathcal{O}(5)$ on $\mathbb{P}^4$. Consider

$$
\begin{array}{ccc}
\overline{M}_{0,1}(\mathbb{P}^4, d) & \xrightarrow{\;\varphi\;} & \mathbb{P}^4 \\
& & \downarrow \pi \\
Z(\widetilde{Q}_d) = \overline{M}_{0,0}(V, d[\mathbb{P}^1]) & \xrightarrow{\;\subset\;} & \overline{M}_{0,0}(\mathbb{P}^4, d)
\end{array}
$$

Here $\widetilde{Q}_d$ is the section of $\mathcal{E}_d := \pi_*(\varphi^* \mathcal{O}(5))$ determined by Q. Note that $\mathcal{E}_d$ is a vector bundle: for $f : C \to \mathbb{P}^4$ a genus-0 stable map, $H^1(C, f^*\mathcal{O}(5)) = 0$. Then look at the number

$$
N_d := \int_{\overline{M}(\mathbb{P}^4,d)} c_{5d+1}(\mathcal{E}_d) \quad :
$$

this ought to count the cardinality of $Z(\widetilde{Q}_d)$, that is the weighted number of rational curves on a quintic threefold. Problem: $N_d \notin \mathbb{Z}$ in general; however, one can see that

$$
N_d = \sum_{k|d} k^{-3} N^o_{d/k}
$$

with N^o_d counting the actual number of rational curves (without contributions of multiple coverings from lower d's, see (3)). This should be an integer.

(3) Multiple coverings of rational curves on Calabi-Yau threefolds.

Consider a $C_0 \cong \mathbb{P}^1$ on a Calabi-Yau threefold V with normal bundle $\mathcal{O}(-1) \oplus \mathcal{O}(-1)$. The space $\overline{M}_{0,0}(V, d[C_0])$ has a connected component $\overline{M}_{0,0}(C_0, d[C_0]) \cong \overline{M}_{0,0}(\mathbb{P}^1, d)$. Its dimension is $(2d-2)$, while the virtual dimension is 0; the obstruction sheaf $\mathcal{F}_d$ is the rank-$(2d-2)$ vector bundle with fiber $H^1(C, f^*(\mathcal{T}_V/\mathcal{T}_{C_0})) \cong \mathbb{C}^2 \otimes H^1(C, f^*(\mathcal{O}(-1)))$ at each point $f : C \to C_0$ (a degree-d stable map). By definition, the contribution of this component will be

$$
M_d := \int_{\overline{M}(\mathbb{P}^1,d)} c_{2d-2}(\mathcal{F}_d) \quad .
$$

Expectation: $M_d = d^{-3}$ (checked by Manin). This explains then the relation between N_d's and N_d^o's.

This relation can be inverted by using the Möbius function:

$$N_d^o = \sum_{k|d} \mu(k) k^{-3} N_{d/k}$$

Goal of the lecture: to explain how Kontsevich ([Kontsevich]) computes in principle the numbers P_d, N_d, M_d. The main tool is *Bott's formula*.

Situation: X is a (smooth) complex projective manifold, $\mathcal{E}$ is a holomorphic vector bundle on X; a complex torus $T \cong \mathbb{C}^* \times \cdots \times \mathbb{C}^*$ acts algebraically on $(X, \mathcal{E})$.

Fact: then X^T is smooth. On each connected component X^γ, $\mathcal{E} = \oplus_{\lambda \in T^\vee} E^{\gamma,\lambda}$, a decomposition into 'eigenbundles' for characters λ. The same for the normal bundles to the fixed loci: for $\mathcal{N}^\gamma := T_X / T_{X^\gamma}$, $\mathcal{N}^\gamma = \oplus_{\lambda \in T^\vee, \lambda \neq 0} \mathcal{N}^{\gamma,\lambda}$.

To write the formula, we introduce the Chern roots of these bundles: e_i for $\mathcal{E}$; $e_i^{\gamma,\lambda}$ for $E^{\gamma,\lambda}$; $n_i^{\gamma,\lambda}$ for $\mathcal{N}^{\gamma,\lambda}$. So $\sum_{k \geq 0} c_k(\mathcal{E}) = \prod_i (1 + e_i)$, and so on.

Now let P be a homogeneous polynomial of degree $\dim X$ in the Chern classes $c_i(\mathcal{E})$ (considered as indeterminates of degree i); that is, a symmetric homogeneous polynomial in the e_i, of degree $\dim X$. Bott's formula then states that

$$\int_X P(e_i) = \sum_\gamma \int_{X^\gamma} \frac{P(e_i^{\gamma,\lambda} + \lambda)}{\prod (n_i^{\gamma,\lambda} + \lambda)} \quad .$$

Every character λ defines a linear form on $\mathrm{Lie}(T)$, so the right-hand-side is a rational function on $\mathrm{Lie}(T)$ (which is constant by the statement, and equals $\int_X P(e_i)$).

We will use Bott's formula on *orbifolds*.

Description of fixed points of the natural action of $T \cong (\mathbb{C}^*)^{n+1}$ on $\overline{M}_{g,k}(\mathbb{P}^n, d)$. This is induced from the action of T on $\mathbb{P}^n$.

Notation: p_i, $i = 1, \ldots, n+1$, are the fixed points of T on $\mathbb{P}^n$. We think of T acting diagonally on $(n+1)$-tuples of homogeneous coordinates, so p_i is the projectivization of the i-th coordinate line in $\mathbb{C}^{n+1}$. For $i \neq j$, let $\ell_{ij} = \ell_{ji}$ be the line in $\mathbb{P}^n$ through p_i and p_j.

Suppose that the stable map $f : C \to \mathbb{P}^n$ represents a point of $\overline{M}_{g,k}(\mathbb{P}^n, d)^T$. Then

(1) $f(C)$ is T-invariant, so a union of lines ℓ_{ij} (points with > 2 nonzero coordinates have ≥ 2-dimensional orbits);

(2) the images of all marked and singular points, and of all contracted components, are points p_i;

(3) a component C^α of V not contracted by f maps onto a line ℓ_{ij}, say with degree d_α. The map $C^\alpha \to \ell_{ij}$ can only be ramified over p_i and p_j; by the Hurwitz formula, necessarily C^α has genus 0 and is totally ramified over p_i and p_j. Note that in particular C^α is smooth; the map is $f(z_1 : z_2) = (0 : \cdots : z_1^{d_\alpha} : \cdots : z_2^{d_\alpha} : \cdots : 0)$.

To each T-fixed $(C; x_1, \ldots, x_k; f)$ we associate a graph Γ (that is, a 1-dimensional finite CW-complex):

(a) Vertices $v \in \text{Vert}(\Gamma)$ correspond to connected components C_v of $f^{-1}(p_i)$ (note: C_v may be a point);

(b) Edges $\alpha \in \text{Edge}(\Gamma)$ correspond to irreducible components C_α (of genus 0) mapping onto lines ℓ_{ij}.

So, the edges α at a vertex v correspond to the non-contracted irreducible components C_α having non-empty intersection with the connected component C_v. The two vertices connected by the edge α corresponding to a C_α, say mapping to ℓ_{ij}, correspond to the unique connected components $C_{v_i,\alpha} \subset f^{-1}(p_i)$, $C_{v_j,\alpha} \subset f^{-1}(p_j)$ resp. which contain $C_\alpha \cap f^{-1}(p_i)$, $C_\alpha \cap f^{-1}(p_j)$ resp. Note that these two vertices are distinct, so Γ has no simple loops.

In short, Γ is made from C by contracting each C_v to a vertex v. In particular, Γ is connected.

(c) Labels on Γ: the vertices v get a number f_v via $f(C_v) = p_{f_v}$; the edges α get labeled by the degree d_α of $f : C_\alpha \to \ell_{ij}$. Further, we define $g_v :=$ the arithmetic genus of the 1-dimensional part of C_v; and $S_v \subset \{1, \ldots, k\}$ to be the set of indices of marked points lying on C_v.

CLAIM. *The connected components of $\overline{M}_{g,k}(\mathbb{P}^n, d)^T$ are naturally labeled by the equivalence classes of connected graphs Γ (with specifications) such that*

(1) *if an edge α connects the vertices u, v, then $f_u \neq f_v$ (in particular, Γ cannot have simple loops);*

(2) $1 - \chi(\Gamma) + \sum_{v \in \text{Vert}(\Gamma)} g_v = g;$

(3) $\sum_{\alpha \in \text{Edge}(\Gamma)} d_\alpha = d;$

(4) $\amalg_v S_v = \{1, \ldots, k\}.$

Kontsevich says (in words): $\overline{M}_{g,k}(\mathbb{P}^n, d)^\Gamma \cong \left(\prod_v \overline{M}_{g_v, \text{val}(v) + \# S_v} \right) / \text{Aut}(\Gamma)$. This isn't quite true, as we will see.

Now for the computation. Assumptions: (1) $g = 0$, so all Γ's are trees, all interior genera $g_v = 0$;

(2) for simplicity, we forget the marked points (for the time being);

(3) notation:

(a) $\overline{M} = \overline{M}(\mathbb{P}^n, d)$ (n, d are fixed);

(b) $[\mathcal{E}]$ for the class in the equivariant K-group with $\mathbb{Q}$-coefficients of a T-equivariant vector bundle $\mathcal{E}$ on $\overline{M}^\Gamma$:

$$K_T^0(\overline{M}^\Gamma) \otimes \mathbb{Q} \cong K^0(\overline{M}^\Gamma) \otimes \mathbb{Q}[T^\vee]$$

(c) $[\chi]$ for the class of a trivial line bundle with T-action given by $\chi \in T^\vee \otimes \mathbb{Q}$.

(4) We systematically decompose fibers of vector bundles as formal linear combinations of other vector spaces, in order to compute the characters. First, the normal bundle:

$$[N_{\overline{M}^\Gamma}] = [T_{\overline{M}}] - [T_{\overline{M}^\Gamma}]$$

$$[T_{\overline{M}}] = [H^0(C, f^*T_{\mathbb{P}^n})] + \sum_{y \in C^\alpha \cap C^\beta, \alpha \neq \beta} [T_y(C^\alpha) \otimes T_y(C^\beta)]$$

$$+ \left(\sum_{y \in C^\alpha \cap C^\beta, \alpha \neq \beta} ([T_y(C^\alpha)] + [T_y(C^\beta)]) - \sum_\alpha [H^0(C^\alpha, T_{C^\alpha})] \right)$$

(Recall: (i) no marked points; (ii) for one component, this is OK; (iii) if we add a component, then we subtract an extra 3-dimensional thing, but we also add 3 dimensions: 1 for smoothing the node, and 2 for moving it. Compare with [F-P].)

$$[T_{\overline{M}^\Gamma}] = 0 + \sum_{y \in C^\alpha \cap C^\beta, \alpha \neq \beta, \alpha, \beta \notin \text{Edge}} [T_y(C^\alpha) \otimes T_y(C^\beta)]$$

$$+ \sum_{y \in C^\alpha \cap C^\beta, \alpha \neq \beta, \alpha \notin \text{Edge}} [T_y(C^\alpha)] - \sum_{\alpha \notin \text{Edge}} [H^0(C^\alpha, T_{C^\alpha})]$$

(Namely, we can only smooth nodes for which both branches are contracted; we can move a node only on a contracted component; and we have infinitesimal automorphisms only on contracted components.) So,

$$[N_{\overline{M}^\Gamma}] = [H^0(C, f^*T_{\mathbb{P}^n})] + [N_{\overline{M}^\Gamma}^{\text{abs}}] \quad , \quad \text{where}$$

$$[N_{\overline{M}^\Gamma}^{\text{abs}}] = \sum_{y; \alpha, \beta \in \text{Edge}} [T_y(C^\alpha) \otimes T_y(C^\beta)] + \sum_{y; \alpha \in \text{Edge}; \beta \notin \text{Edge}} [T_y(C^\alpha) \otimes T_y(C^\beta)]$$

$$+ \left(\sum_{y; \alpha \in \text{Edge}} [T_y(C^\alpha)] - \sum_{\alpha \in \text{Edge}} [H^0(C^\alpha, T_{C^\alpha})] \right)$$

The only nontrivial vector bundle term: $T_y(C^\beta)$ for β not an edge, in the second summand. However, this term has trivial character (which will simplify the computation).

More notions and notations. A *flag* is an edge α with arrow: (v, α), $v \in \text{Vert}(\Gamma)$ thought of as the source of the arrow. This is unambiguous, as Γ has no simple loops. For $F = (v, \alpha)$ a flag, the *weight* of F is

$$w_F := \frac{(\lambda_{f_v} - \lambda_{f_u})}{d_\alpha}$$

where u is the other vertex of α. Here λ_i, $i = 1, \ldots, n+1$, are the natural coordinates on $\text{Lie}(T)$; w_F is the character of T for the action on $T_{C^\alpha, C_v \cap C^\alpha}$. If $\overline{F} = (u, \alpha)$ is the dual flag then of course $w_{\overline{F}} = -w_F$.

We need the following integral:

$$I(w_1, \ldots w_k) := \int_{\overline{M}_{0,k}} \frac{1}{w_i + c_1(T_{x_i}(C))}$$

$$= \sum_{d_i \geq 0, \sum d_i = k-3} \prod_{i=1}^{k} w_i^{-d_i-1} \underbrace{\langle \tau_{d_1} \cdots \tau_{d_k} \rangle}_{= \frac{(k-3)!}{d_1! \cdots d_k!}} \qquad \text{(Witten's notation)}$$

$$= (\sum w_i^{-1})^{k-3} (\prod w_i^{-1})$$

Recall that we are to compute integrals $\int_{M'} \dfrac{P(e_i^{\Gamma,\lambda} + \lambda)}{\prod_i (n_i^{\Gamma,\lambda} + \lambda)}$. We will see
that in all three examples the equivariant vector bundles are trivial as vector
bundles: $e_i^{\Gamma,\lambda} = 0$ always. So $P(\cdots)$ is just a constant for the integral. Similarly
for many of the $n_i^{\Gamma,\lambda}$, but not for all: exactly the terms

$$\sum_{y; \alpha \in \text{Edge}, \beta \notin \text{Edge}} [T_y(C^\alpha) \otimes T_y(C^\beta)]$$

in $[N]$ (written additively) are nontrivial bundles. Writing multiplicatively, and
putting it in the denominator, we get

$$\int_{\overline{M}^\Gamma} \frac{1}{\prod_{y; \alpha \in \text{Edge}, \beta \notin \text{Edge}} (w_F + c_1(T_y(C^\beta)))}$$

(with $F = (y, \alpha)$) as the only part that needs to be integrated.

Now $\overline{M}^\Gamma$ "$=$" $\left(\prod_v \overline{M}_{0,\text{val}(v)+\#S_v} \right) / \text{Aut}(\Gamma) \mapsto \prod_{v, \text{val}(v) \geq 3} \overline{M}_{0,\text{val}(v)}$, forgetting marked points and the action of $\text{Aut}(\Gamma)$. So this becomes

$$\prod_{v, \text{val}(v) \geq 3} \left\{ \left(\sum_{F=(v,\alpha)} w_F^{-1} \right)^{\text{val}(v)-3} \prod_{F=(v,\alpha)} w_F^{-1} \right\} =: \prod_{v, \text{val}(v) \geq 3} (*)$$

by the integral computed above.

Now two remarkable steps recover the vertices of lower valence:

(1) $\sum_{y; \alpha, \beta \in \text{Edge}} [T_y(C^\alpha) \otimes T_y(C^\beta)]$ corresponds to the vertices with va-
lency 2. The contribution (to the denominator) is $\prod_{v, \text{val}(v)=2} (w_{F_1(v)} + w_{F_2(v)})^{-1}$. Since $(\frac{1}{a} + \frac{1}{b})^{-1} \frac{1}{a} \frac{1}{b} = \frac{1}{a+b}$, this gives exactly

$$\prod_{v, \text{val}(v)=2} (*)$$

(2) $-\sum_{\alpha\in\text{Edge}}[H^0(C^\alpha, T_{C^\alpha})] = -\sum_{\alpha\in\text{Edge}}([-w_{F(\alpha)}] + [0] + [w_{F(\alpha)}])$

$= -\sum_{\text{flags } F}[w_F] - (\#\text{edges})[0]$. On the other hand, $\sum_{y,\alpha\in\text{Edge}}[T_y(C^\alpha)]$

$= \sum_{F=(v,\alpha),\text{val}(v)\geq 2}[w_F]$. All in all, we get

$$-\sum_{F=(v,\alpha),\text{val}(v)=1}[w_F] - (\#\text{edges})[0]$$

The term $-(\#\text{edges})[0]$ will cancel out. The rest gives $\prod w_F$, and after all $w_F = (w_F^{-1})((w_F)^{-1})^{-2}$; putting all together, the contribution of $[N_M^{abs}] + (\#\text{edges})[0]$ is

(A)
$$\prod_v \left\{ \left(\sum_{F=(v,\alpha)} w_F^{-1} \right)^{\text{val}(v)-3} \prod_{F=(v,\alpha)} w_F^{-1} \right\}$$

Next, we examine $[H^0(C, f^*T_{\mathbb{P}^n})]$. We have a short exact sequence

$$0 \to H^0(C, f^*T_{\mathbb{P}^n}) \to \oplus_{\alpha\in\text{Edge}} H^0(C^\alpha, f^*T_{\mathbb{P}^n}) \to \oplus_v T_{p_{f_v}}\mathbb{P}^n \otimes \mathbb{C}^{\text{val}(v)-1} \to 0$$

(global sections are tuples of sections over edges, that agree at each v for all edges at v). To see what the middle term is, recall that f on C_α is given via

$$X_i(f(z)) = z_1^{d_\alpha}, X_j(f(z)) = z_2^{d_\alpha}, X_k(f(z)) = 0 \quad \forall k \neq i,j$$

for $z = (z_1 : z_2)$ a coordinate on $\mathbb{P}^1 \cong C^\alpha$ mapping $d_\alpha : 1$ onto ℓ_{ij}, and the X_i homogeneous coordinates on $\mathbb{P}^n$.

A calculation shows that the following elements form a basis for $H^0(C^\alpha, f^*T_{\mathbb{P}^n})$:

(1) $z^a X_i \frac{\partial}{\partial X_i}$ $-d_\alpha \leq a \leq d_\alpha$;

(2) $z_1^a z_2^b \frac{\partial}{\partial X_k}$ $a + b = d_\alpha, 0 \leq a, b; k \neq i, j (k \in \{1, \dots, n+1\})$

There is a unique element with trivial T-action: $X_i \frac{\partial}{\partial X_i}$, yielding $(\#\text{edges})[0]$ and giving the promised cancellation.

Now z corresponds to $\frac{\lambda_i - \lambda_j}{d_\alpha} = w_F$ ($F = (v, \alpha)$ with $f(C_V) = p_i$); X_k corresponds to λ_k, z_1 to λ_i/d_α, and z_2 to λ_j/d_α.

The third term in the short exact sequence gives

$$(1 - \text{val}(v)) \sum_{j\neq f_v} [\lambda_{f_v} - \lambda_j]$$

since the elements $X_{f_v}\frac{\partial}{\partial X_j}$ with $j \neq f_v$ form a basis of $T_{p_{f_v}}\mathbb{P}^n$. In total, we find

(B)
$$\prod_{\alpha \text{ joining } v_1,v_2} \left\{ \frac{\left(\frac{d_\alpha}{\lambda_{f_{v_1}} - \lambda_{f_{v_2}}}\right)^{2d_\alpha}}{(-1)^{d_\alpha}((d_\alpha)!)^2} \prod_{k\neq f_{v_1}, k\neq f_{v_2}} \prod_{a+b=d_\alpha; a,b\geq 0} \frac{1}{\frac{a}{d_\alpha}\lambda_{f_{v_1}} + \frac{b}{d_\alpha}\lambda_{f_{v_2}} - \lambda_k} \right\}$$

$$\cdot \prod_v \left(\prod_{j\neq f_v} (\lambda_{f_v} - \lambda_j) \right)^{\text{val}(v)-1}$$

Marked points. Claim: the only effect of allowing marked points is that in term (A), val(v) must be replaced by val(v) + #S_v. Also, the graphs considered must allow extra tails, corresponding to the markings.

The last contribution (C) we need to consider is the contribution of the vector bundles $\mathcal{E}$ in the three examples: trivial bundles, twisted with suitable characters.

(1) $\prod_v (\lambda_{f_v})^{2\#S_v}$;

(2) The exact sequence

$$0 \to H^0(C, f^*\mathcal{O}(5)) \to \oplus_{\alpha \in \text{Edge}} H^0(C^\alpha, f^*\mathcal{O}(5)) \to \oplus_v \mathcal{O}(5)_{p_{f_v}} \otimes \mathbb{C}^{\text{val}(v)-1} \to 0$$

is used to obtain

$$\prod_{\alpha \text{ joining } v_1, v_2} \left(\prod_{a,b \geq 0, a+b=5d_\alpha} \frac{a\lambda_{f_{v_1}} + b\lambda_{f_{v_2}}}{d_\alpha} \right) \prod_v (5\lambda_{f_v})^{1-\text{val}(v)}$$

(3) Similarly,

$$0 \to \oplus_v (\mathcal{O}(-1)_{p_{f_v}} \otimes \mathbb{C}^{\text{val}(v)-1}) \to H^1(C, f^*\mathcal{O}(-1))$$
$$\to \oplus_{\alpha \in \text{Edge}} H^1(C^\alpha, f^*\mathcal{O}(-1)) \to 0$$

yields

$$\left\{ \prod_{\alpha \text{ joining } v_1, v_2} \left(\prod_{a,b<0, a+b=-d_\alpha} \frac{a\lambda_{f_{v_1}} + b\lambda_{f_{v_2}}}{d_\alpha} \right) \cdot \prod_v (-\lambda_{f_v})^{\text{val}(v)-1} \right\}^2$$

The final sum is

$$\sum_{\hat{\Gamma}} \frac{1}{\text{Aut}(\hat{\Gamma})} (A)(B)(C)$$

where $\hat{\Gamma}$ are the graphs with tails (for marked points), and as mentioned above (A) has val(v) + #S_v instead of val(v) if there are marked points.

This is what Kontsevich writes.

One hitch: we are doing integrals over orbifolds, so we should not forget automorphisms of general elements (maps). The order of this group is (#Aut($\hat{\Gamma}$)) · ($\prod_{\alpha \in \text{Edge}} d_\alpha$); we then need to divide by this.

Moral. If you can formulate your favorite counting problem as the computation of the degree of a Chern class on a Kontsevich space $\overline{M}_{g,k}(\mathbb{P}^n, d)$, you have a good chance of reducing it to a sum over graphs. This should at least enable you to calculate the first few cases by computer. Also, physicists and combinatorists have tricks to do sums over trees and graphs.

Examples. First example, $d = 1$. Since $d_\alpha \geq 1$ and $\sum_\alpha d_\alpha = d = 1$, we can only have one edge. There are 4 ways to distribute the 2 marked points, and 3 different labelings: $(1)(2)$, $(1)(3)$, $(2)(3)$. For $(1)(2)$: let w_1 be the flag at 1, etc. So $w_1 = \lambda_1 - \lambda_2$, $w_2 = \lambda_2 - \lambda_1$, and

(A)$= (w_1^{-1})^{\#S_1-1}(w_2^{-1})^{\#S_2-1} = (-1)^{\#S_2-1}$;

(B)$= \frac{-1}{(\lambda_1-\lambda_2)^2}\frac{1}{\lambda_1-\lambda_3}\frac{1}{\lambda_2-\lambda_3}$ (all valencies are 1);

(C)$= (\lambda_1)^{2\#S_1}(\lambda_2)^{2\#S_2}$.

Total for $(1)(2)$:

$$\frac{1}{(\lambda_1 - \lambda_2)^2(\lambda_1 - \lambda_3)(\lambda_2 - \lambda_3)}(\lambda_1^4 + \lambda_2^4 - 2\lambda_1^2\lambda_2^2) = \frac{(\lambda_1 + \lambda_2)^2}{(\lambda_1 - \lambda_3)(\lambda_2 - \lambda_3)}$$

Now take the corresponding terms for $(1)(3)$ and $(2)(3)$, and discover the nice identity

$$\frac{(\lambda_1 + \lambda_2)^2}{(\lambda_1 - \lambda_3)(\lambda_2 - \lambda_3)} + \frac{(\lambda_1 + \lambda_3)^2}{(\lambda_1 - \lambda_2)(\lambda_3 - \lambda_2)} + \frac{(\lambda_2 + \lambda_3)^2}{(\lambda_2 - \lambda_1)(\lambda_3 - \lambda_1)} = 1 \quad ,$$

the number of lines through 2 points.

The second example, for $d = 1$, also involves trees with a single edge. Carefully evaluating each term gives the total

$$\frac{25\,\lambda_1\,\lambda_2\,(4\,\lambda_1 + \lambda_2)\,(3\,\lambda_1 + 2\,\lambda_2)\,(2\,\lambda_1 + 3\,\lambda_2)\,(\lambda_1 + 4\,\lambda_2)}{(\lambda_1 - \lambda_3)\,(\lambda_2 - \lambda_3)\,(\lambda_1 - \lambda_4)\,(\lambda_2 - \lambda_4)\,(\lambda_1 - \lambda_5)\,(\lambda_2 - \lambda_5)}$$
$$+ \frac{25\,\lambda_1\,\lambda_3\,(4\,\lambda_1 + \lambda_3)\,(3\,\lambda_1 + 2\,\lambda_3)\,(2\,\lambda_1 + 3\,\lambda_3)\,(\lambda_1 + 4\,\lambda_3)}{(\lambda_1 - \lambda_2)\,(-\lambda_2 + \lambda_3)\,(\lambda_1 - \lambda_4)\,(\lambda_3 - \lambda_4)\,(\lambda_1 - \lambda_5)\,(\lambda_3 - \lambda_5)}$$
$$+ \frac{25\,\lambda_2\,\lambda_3\,(4\,\lambda_2 + \lambda_3)\,(3\,\lambda_2 + 2\,\lambda_3)\,(2\,\lambda_2 + 3\,\lambda_3)\,(\lambda_2 + 4\,\lambda_3)}{(-\lambda_1 + \lambda_2)\,(-\lambda_1 + \lambda_3)\,(\lambda_2 - \lambda_4)\,(\lambda_3 - \lambda_4)\,(\lambda_2 - \lambda_5)\,(\lambda_3 - \lambda_5)}$$
$$+ \frac{25\,\lambda_1\,\lambda_4\,(4\,\lambda_1 + \lambda_4)\,(3\,\lambda_1 + 2\,\lambda_4)\,(2\,\lambda_1 + 3\,\lambda_4)\,(\lambda_1 + 4\,\lambda_4)}{(\lambda_1 - \lambda_2)\,(\lambda_1 - \lambda_3)\,(-\lambda_2 + \lambda_4)\,(-\lambda_3 + \lambda_4)\,(\lambda_1 - \lambda_5)\,(\lambda_4 - \lambda_5)}$$
$$+ \frac{25\,\lambda_2\,\lambda_4\,(4\,\lambda_2 + \lambda_4)\,(3\,\lambda_2 + 2\,\lambda_4)\,(2\,\lambda_2 + 3\,\lambda_4)\,(\lambda_2 + 4\,\lambda_4)}{(-\lambda_1 + \lambda_2)\,(\lambda_2 - \lambda_3)\,(-\lambda_1 + \lambda_4)\,(-\lambda_3 + \lambda_4)\,(\lambda_2 - \lambda_5)\,(\lambda_4 - \lambda_5)}$$
$$+ \frac{25\,\lambda_3\,\lambda_4\,(4\,\lambda_3 + \lambda_4)\,(3\,\lambda_3 + 2\,\lambda_4)\,(2\,\lambda_3 + 3\,\lambda_4)\,(\lambda_3 + 4\,\lambda_4)}{(-\lambda_1 + \lambda_3)\,(-\lambda_2 + \lambda_3)\,(-\lambda_1 + \lambda_4)\,(-\lambda_2 + \lambda_4)\,(\lambda_3 - \lambda_5)\,(\lambda_4 - \lambda_5)}$$
$$+ \frac{25\,\lambda_1\,\lambda_5\,(4\,\lambda_1 + \lambda_5)\,(3\,\lambda_1 + 2\,\lambda_5)\,(2\,\lambda_1 + 3\,\lambda_5)\,(\lambda_1 + 4\,\lambda_5)}{(\lambda_1 - \lambda_2)\,(\lambda_1 - \lambda_3)\,(\lambda_1 - \lambda_4)\,(-\lambda_2 + \lambda_5)\,(-\lambda_3 + \lambda_5)\,(-\lambda_4 + \lambda_5)}$$
$$+ \frac{25\,\lambda_2\,\lambda_5\,(4\,\lambda_2 + \lambda_5)\,(3\,\lambda_2 + 2\,\lambda_5)\,(2\,\lambda_2 + 3\,\lambda_5)\,(\lambda_2 + 4\,\lambda_5)}{(-\lambda_1 + \lambda_2)\,(\lambda_2 - \lambda_3)\,(\lambda_2 - \lambda_4)\,(-\lambda_1 + \lambda_5)\,(-\lambda_3 + \lambda_5)\,(-\lambda_4 + \lambda_5)}$$
$$+ \frac{25\,\lambda_3\,\lambda_5\,(4\,\lambda_3 + \lambda_5)\,(3\,\lambda_3 + 2\,\lambda_5)\,(2\,\lambda_3 + 3\,\lambda_5)\,(\lambda_3 + 4\,\lambda_5)}{(-\lambda_1 + \lambda_3)\,(-\lambda_2 + \lambda_3)\,(\lambda_3 - \lambda_4)\,(-\lambda_1 + \lambda_5)\,(-\lambda_2 + \lambda_5)\,(-\lambda_4 + \lambda_5)}$$
$$+ \frac{25\,\lambda_4\,\lambda_5\,(4\,\lambda_4 + \lambda_5)\,(3\,\lambda_4 + 2\,\lambda_5)\,(2\,\lambda_4 + 3\,\lambda_5)\,(\lambda_4 + 4\,\lambda_5)}{(-\lambda_1 + \lambda_4)\,(-\lambda_2 + \lambda_4)\,(-\lambda_3 + \lambda_4)\,(-\lambda_1 + \lambda_5)\,(-\lambda_2 + \lambda_5)\,(-\lambda_3 + \lambda_5)}$$

which simplifies to 2875, the number of lines on a quintic threefold.

11. Equivariant cohomology—P. Belorousski
December 12, 1996

The basic reference is [A-B]; also, see [Ginzburg].

General theory. Let G be a compact connected Lie group, and consider its universal principal bundle $EG \to BG$. Here EG is a contractible space, with a *free* (right) G-action (this defines it uniquely up to homotopy), and $BG = EG/G$. One often works with finite–dimensional approximations BG_N, EG_N (which can be chosen smooth and compact).

Let X be a smooth manifold with a smooth left G-action. The following is what is known as the *Borel's mixing construction*. Consider $EG \times X$ (on which G acts freely by $(\alpha, x) \overset{g}{\to} (\alpha g^{-1}, gx)$) and the quotients:

$$
\begin{array}{ccccc}
EG & \longleftarrow & EG \times X & \longrightarrow & X \\
\downarrow & & \downarrow & & \downarrow \\
BG & \overset{\pi}{\longleftarrow} & (EG \times X)/G & \overset{\sigma}{\longrightarrow} & X/G
\end{array}
$$

$X_G := (EG \times X)/G$ is the *homotopy quotient* of X by G.

DEFINITION. The *equivariant cohomology* of X is $H_G^*(X) := H^*(X_G)$.

The map $\pi : X_G \to BG$ is a bundle with fiber X; in the associated Leray spectral sequence, $E_2^{p,q} = H^p(BG, H^q(X)) \implies H^{p+q}(X_G)$. In our case, BG will be simply connected and we will work over $\mathbb{C}$, so by Künneth $H^p(BG, H^q(X)) \cong H^p(BG) \otimes H^q(X)$.

Also, we have a pull-back map $\pi^* : H^*(BG) \to H_G^*(X)$, which gives $H_G^*(X)$ an $H^*(BG)$-module structure. We will write H_G^* for the coefficient ring $H^*(BG) = H_G^*(\mathrm{pt})$.

The map σ is not a fibration. For x a point in X, O_x its orbit, $\sigma^{-1}(O_x) = EG/G_x = BG_x$, where G_x is the stabilizer of x. If G acts freely on X, then σ is a homotopy equivalence, and $H_G^*(X) \cong H^*(X/G)$. In this sense, equivariant cohomology can say something new only if the action of G on X is not free to start with.

We also have the fiber inclusion over the base point of BG, $i : X \hookrightarrow X_G$, and the corresponding pull-back map $i^* : H_G^*(X) \to H^*(X)$. Finally, we have the push-forward $\pi_* : H_G^*(X) \to H_G^{*-n}$; and a pairing

$$
\langle \cdot, \cdot \rangle : H_G^i(X) \otimes H_G^j(X) \to H_G^{i+j-n}
$$

with $n = \dim X$.

We can say a lot more in the case of the *Hamiltonian actions* on *symplectic manifolds*.

Reminder. A smooth manifold X^{2n} is *symplectic* if it is endowed with a form $\omega \in \Omega^2(X)$ such that

(1) ω is closed;
(2) the symplectic volume form $\omega^n/n!$ is nowhere 0 on X.

This gives $C^\infty(X)$ a Lie algebra structure, with the Poisson bracket $\{f, g\} = df(g)$ as the Lie bracket, after identifying df with a vector field via ω. $C^\infty(X)$ is an extension of the Lie subalgebra $\mathrm{Vect}_\omega(X) \subset \mathrm{Vect}(X)$ of vector fields preserving ω.

An action of a group G on X is *symplectic* if it preserves the form ω. We can ask more: the action of G on X gives a homomorphism of Lie algebras $\mathfrak{g} \to \mathrm{Vect}(X)$, and we can require that there be a lifting

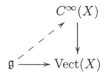

If such a lifting exists, the action is called *Hamiltonian* (or *Poisson*).

THEOREM. *If X has a Hamiltonian action of G, then the Leray spectral sequence degenerates at the second term. As a consequence, $H_G^*(X)$ is a free H_G^*-module, isomorphic to $H^*(X) \otimes H_G^*$.*

Also, the restriction map $i^* : H_G^*(X) \to H^*(X)$ is surjective. Furthermore, the pairing $\langle \cdot, \cdot \rangle$ is non-degenerate, so we have an isomorphism

$$H_G^*(X) \xrightarrow{\sim} \mathrm{Hom}_{H_G^*}(H_G^*(X), H_G^*)$$

Thus if the action is Hamiltonian we may pick a basis $\{h_1, \ldots, h_m\}$ of the free H_G^*-module $H_G^*(X)$, and get that $\det((\langle h_i, h_j \rangle)) \in H_G^0 \cong \mathbb{C}$ (since the pairing is nondegenerate, the determinant must be invertible).

We will look at algebraic actions of reductive groups. Fact: they are all Hamiltonian.

Classifying spaces. First, take $G = \mathbb{C}^*$. Then $EG_N = \mathbb{C}^{N+1} \setminus \{0\}$, and $EG_N \to BG_N$ is simply the factor map $\mathbb{C}^{N+1} \setminus \{0\} \to \mathbb{P}^N$. In the limit, $EG \to BG$ is the map $\mathbb{C}^\infty \setminus \{0\} \to \mathbb{P}^\infty$.

We can think of EG as the complement to the zero section in the total space of $\mathcal{O}(-1)$ over $\mathbb{P}^\infty$. We get $H_{\mathbb{C}^*}^* = H^*(\mathbb{P}^\infty) = \mathbb{C}[\hbar]$, where $\deg \hbar = 2$, $\hbar = c_1(\mathcal{O}(1))$.

For $G = T^n = (\mathbb{C}^*)^n$, $EG = (\mathbb{C}^\infty \setminus \{0\})^n$ and $BG = (\mathbb{P}^\infty)^n$. The coefficient ring is $H_{T^n}^* = \mathbb{C}[\hbar_1, \ldots, \hbar_n]$, with $\hbar_i = c_1(\mathcal{O}_i(1))$.

For $G = \mathrm{GL}(n)$ the finite dimensional approximation is $EG_N = \mathrm{Fr}(n, N)$, the bundle of n-frames in $\mathbb{C}^n$ over the Grassmannian $\mathrm{Gr}(n, N) = BG_N$. In the limit, $EG \to BG$ is $\mathrm{Fr}(n, \infty) \to \mathrm{Gr}(n, \infty)$. $H_G^* = \mathbb{C}[s_1, \ldots, s_n]$, where $s_i = c_i(\mathcal{S}^n)$, with $\mathcal{S}^n$ the universal subbundle over $\mathrm{Gr}(n, \infty)$. Of course $\deg s_i = 2i$. We can think of the s_i's as the elementary symmetric functions (up to sign) in the $\hbar_i$'s:

$$\mathbb{C}[s_1, \ldots, s_n] = \mathbb{C}[\hbar_1, \ldots, \hbar_n]^{S_n}, \qquad s_i = (-1)^i s_i(\hbar).$$

Important example. $T = T^{n+1} = (\mathbb{C}^*)^{n+1} \hookrightarrow \mathrm{GL}(n+1)$ acting on $\mathbb{P}^n$:

$$(t_0, \ldots, t_n) \mapsto \begin{pmatrix} t_0 & \cdots & 0 \\ & \ddots & \\ 0 & \cdots & t_n \end{pmatrix}$$

Here EG sits as the complement to the zero section in $E = \mathcal{O}_0(-1) \oplus \cdots \oplus \mathcal{O}_n(-1)$ over $(\mathbb{P}^\infty)^{n+1}$. The homotopy quotient $\mathbb{P}^n_T$ is $\mathbb{P}E$. So

$$H^*_T(\mathbb{P}^n) = \mathbb{C}[\hbar_0, \ldots, \hbar_n; y]/(y^{n+1} + y^n c_1(E) + \cdots + c_{n+1}(E))$$
$$= \mathbb{C}[\hbar_0, \ldots, \hbar_n; y]/(y - \hbar_0) \cdots (y - \hbar_n)$$

The base ring here is $\mathbb{C}[\hbar_0, \ldots, \hbar_n]$ as seen above, and the module structure is the one obvious from the presentation.

Next note that this action has $(n+1)$ fixed points: $p_k = (0 : \cdots : \overset{k}{1} : \cdots : 0)$. The inclusion $i_k : p_k \hookrightarrow \mathbb{P}^n$ is equivariant, so we have a push-forward on $H^*_T()$:

$$(i_k)_* : \mathbb{C}[\hbar_0, \ldots, \hbar_n] \to \mathbb{C}[\hbar_0, \ldots, \hbar_n; y]/(y - \hbar_0) \cdots (y - \hbar_n)$$

What is this map? It corresponds to the embedding $\mathbb{P}(\mathcal{O}_k(-1)) \hookrightarrow \mathbb{P}(E)$ of the projectivisation of the k-th copy of $\mathcal{O}(-1)$ over $(\mathbb{P}^\infty)^{n+1}$. Now if

$$0 \to S \to E \to Q \to 0$$

is an exact sequence of bundles over X, then $\mathbb{P}(S) \hookrightarrow \mathbb{P}(E)$ is the zero locus of a section of $p^* Q \otimes \mathcal{O}_{\mathbb{P}(E)}(1)$, where p denotes the projection $\mathbb{P}(E) \to X$. So, $[\mathbb{P}(S)] = c_{top}(p^* Q \otimes \mathcal{O}_{\mathbb{P}(E)}(1))$.

Using this, it is straightforward to compute the class of $\mathbb{P}(\mathcal{O}_k(-1))$ and find that under $(i_k)_*$

$$1 \mapsto (y - \hbar_0) \cdots \widehat{(y - \hbar_k)} \cdots (y - \hbar_n)$$

What about the pull-back

$$(i_k)^* : \mathbb{C}[\hbar_0, \ldots, \hbar_n; y]/(y - \hbar_0) \cdots (y - \hbar_n) \to \mathbb{C}[\hbar_0, \ldots, \hbar_n] \qquad ?$$

Here $\hbar_i \mapsto \hbar_i$ for all i; as for the image of y, the $\mathcal{O}_{\mathbb{P}(E)}(1)$ restricts to $\mathcal{O}(1)$ on $\mathbb{P}(\mathcal{O}_k(-1))$. This says that $y \mapsto \hbar_k$.

The composition

$$(i_k)^* (i_k)_* : \mathbb{C}[\hbar_0, \ldots, \hbar_n] \to \mathbb{C}[\hbar_0, \ldots, \hbar_n]$$

is given by

$$1 \mapsto (\hbar_k - \hbar_0) \cdots \widehat{(\hbar_k - \hbar_k)} \cdots (\hbar_k - \hbar_n)$$

Remark: this is precisely the equivariant top Chern class of the (equivariant) normal bundle of p_k in $\mathbb{P}^n$.

Localization. Let T be a torus acting on X (assume that we are in the algebraic situation for simplicity), and let X^T be the fixed point locus. The inclusion $X^T \subset X$ is equivariant, so we have a H_T^*-module restriction homomorphism $H_T^*(X) \to H_T^*(X^T)$.

CLAIM. *This map is injective, and its cokernel is H_T^*-torsion.*

Hence, it is an isomorphism after localization at a suitable $f \in H_T^*$. Since T acts trivially on X^T, we get $H_T^*(X^T) \cong H^*(X^T) \otimes H_T^*$, and hence

$$H_T^*(X) \cong H^*(X^T) \otimes H_T^*$$

after a suitable localization.

Appendix: The computer program farsta—A. Kresch

I. Purpose

[DF-I]: Look at examples.

For simple varieties,

Given associativity relations + several N's,

derive more N's.

farsta: Automate this process.

To obtain farsta

Go to http://www.math.uchicago.edu/~kresch
follow the link to farsta
uncompress, untar, and compile by typing make

Documentation (view with more, print with lpr)

farsta-documentation
farsta-examples

[F-P]: X, $\operatorname{rk} H^* X = m + 1$

Get $(m^4 - 2m^3 + 3m^2 - 2m)/8$ associativity relations;

each is an equation of formal power series.

Isolating coefficients, each associativity relation yields a family of relations in N's.

What farsta can do:

1. From description of $H^* X$ and K_X, derive associativity relations.

2. Given associativity relation and particular degree,

derive equation among N's.

3. Substitute known numbers; solve if equation reduces to just one unknown.

4. Store linear relations, and do linear algebra to solve for N's.

II. Example

$$X = \mathbb{P}^3$$
$$H^*X = \mathbb{Z}\langle T_0, T_1, T_2, T_3 \rangle$$
$$-K_X = 4T_1$$

Relations:

6 relations $R1$–$R6$.

E.g.

$(R2)$ $\underset{2}{\overset{1}{}}\!\!\asymp\!\!\underset{2}{\overset{3}{}}$ $\Gamma_{233} = \Gamma_{113}\Gamma_{222} - \Gamma_{112}\Gamma_{223}$

where

$$\Gamma = \sum_{a+2b=4c} N(c; a, b)\, e^{cy_1} \frac{y_2^a}{a!} \frac{y_3^b}{b!}.$$

Equating coefficients of $e^{2y_1} y_2 y_3$ yields

$$N(2; 2, 3) = N(1; 0, 2)N(1; 4, 0) - N(1; 2, 1)^2.$$

Same relation, coefficient of $e^{2y_1} y_2^3$:

$$\frac{1}{6}N(2; 4, 2) = \frac{1}{3}N(1; 2, 1)N(1; 4, 0).$$

From these equations, the numbers

$$N(1; 0, 2) = 1$$
$$N(1; 2, 1) = 1$$
$$N(1; 4, 0) = 2$$

determine

$$N(2; 2, 3) = 1$$
$$N(2; 4, 2) = 4.$$

Computer:

```
>A P3
dim(P3) = ?              >3
basis H^*(P3) = ?    >y0 y1 y2 y3
rank H^2(P3) = ?     >1
     (complex) codim(y2) = ?              >2
     (complex) codim(y3) = ?              >3
enter classical potential function:
(1/2) y0 2 y3 + y0 y1 y2 + (1/6) y1 3
rank Pic(P3) = ?     >1
     PD(B_1)  in H^2(P3) = ?              >y1
TP3 = tangent bundle on P3
     int_(B_1)  c_1(TP3) = ?              >4

>R 2
(2) (2,3,3,4) 1 G_y2,y3,y3 + 1 G_y1,y1,y2
   G_y2,y2,y3 - 1 G_y1,y1,y3 G_y2,y2,y2 = 0

>N 1 0 2 1
>N 1 2 1 1
>N 1 4 0 2
>E 2 2 1 1
Relation (2)   2 ; 1 1  gives  N(2;2,3) = 1
>E 2 2 3 0
Relation (2)   2 ; 3 0  gives  N(2;4,2) = 4
```

III. Exhaustive search

                try lots of E's

    >A P3

         .

         .   (enter description of $\mathbb{P}^3$).

         .

    >N 1 0 2 1

    >X 1 6 1 4

    ... tells the computer to try exhaustively:

   * relations 1 through 6, (i.e., all)

   * curve class $1T_1$ through $4T_1$.

```
Relation (1)   1 ; 0 0   gives   N(1;2,1) = 1
Relation (4)   1 ; 1 0   gives   N(1;4,0) = 2
Relation (2)   2 ; 3 0   gives   N(2;4,2) = 4
Relation (3)   2 ; 2 0   gives   N(2;2,3) = 1
Relation (3)   2 ; 0 1   gives   N(2;0,4) = 0
Relation (4)   2 ; 3 1   gives   N(2;6,1) = 18
Relation (6)   3 ; 5 0   gives   N(2;8,0) = 92
Relation (2)   3 ; 5 1   gives   N(3;6,3) = 190

     ...

Relation (3)   4 ; 0 5   gives   N(4;0,8) = 4
```

IV. Linear algebra:

    >L1

tells the computer to keep linear relationships among N's in memory.

Recursively substitutes and backsubstitutes to determine more N's
Necessary, e.g., for $G(2,4)$.

V. Observations:

1. Specifying cone of effective classes is unimportant —

    farsta figures it out and gives zeros for noneffective β (even w/o any input!).

2. Can change canonical class to get new solutions.

3. Can work with H^* of an orbifold — any ring with $\mathbb{Q}$-Poincaré duality works.

4. Choice of basis can have big impact on performance.

VI. Performance (on a Sun SparcSTATION 20):

* $G(2,4)$ $m = 5$ 55 relations
                $deg = 1, 2, 3,$ 4   in 90 seconds.
                              —all— most

* Hilb$_2$ $\mathbb{P}^2$ $m = 8$ 406 relations
                $(0,0) \leq (d_1, d_2) \leq (2,2)$ in 3 minutes;
                $(0,0) \leq (d_1, d_2) \leq (2,4)$ in 10 hours;
                $(0,0) \leq (d_1, d_2) \leq (4,4)$ in 45 hours.

* $G(2,5)$ $m = 9$ 666 relations
                              — basis of Schubert cycles
            89 out of 139 $d = 1$ N's in 3 seconds;
            775 out of 865 $d = 2$ N's in 5 minutes;
            3478 out of 3608 $d = 3$ N's in 4 hours;
            11512 out of 11682 $d = 4$ N's in 104 hours.

* $G(2,5)$ $m = 9$ — Tom Graber's basis
            About 10 percent faster.

* $G(2,6)$ $m = 14$ 4186 relations
            573 out of 877 $d = 1$ N's in 6 minutes;
            7675 out of 8820 $d = 2$ N's in 36 hours.

* $G(3,6)$ $m = 19$ 14706 relations
            Can't even get started.

VII. Limitations

* No odd cohomology.

* No free parameters —

numbers only (rational, multi-precision numerator and denominator).

* Memory limitations —

Linear expression buffer tends to fill up.

* Practical limit on $m = \operatorname{rk} H^* X - 1$:

$$m < 10 \qquad\qquad \text{usually fine;}$$
$$10 \leq m \leq 15 \qquad\qquad \text{questionable;}$$
$$m > 15 \qquad\qquad \text{probably hopeless.}$$

References

[A-B]   M. Atiyah, R. Bott, *The moment map and equivariant cohomology*, Topology **23** (1984), 1–28.

[Behrend]   K. Behrend, *Gromov-Witten invariants in algebraic geometry*, 9601011 (1996).

[B-F]   K. Behrend, B. Fantechi, *The intrinsic normal cone*, alg-geom/9601010(1996).

[B-M]   K. Behrend, Y. Manin, *Stacks of stable maps and Gromov-Witten invariants*, Duke Math. J. **85** (1996), 1–60.

[Bertram]   A. Bertram, *Quantum Schubert calculus*, preprint (1996).

[B-V]   J. M. Boardman, R. M. Vogt, *Homotopy invariant algebraic structures on topological spaces*, LNM 347, Springer, 1973.

[CDGP]   P. Candelas, X. de la Ossa, P. Green, L. Parkes, *A pair of Calabi-Yau manifolds as an exactly soluble superconformal theory*, Nucl. Phys. B **359** (1991), 21–74.

[Choi]   Y. Choi, *Severi degrees in cogenus 4*, alg-geom/9601013 (1996).

[DF-I]   P. Di Francesco, C. Itzykson, *Quantum intersection rings*, The moduli space of curves, (Dijkgraaf, Faber, van der Geer), Birkhäuser, 1995, pp. 81–148.

[Dubrovin]   B. Dubrovin, *The geometry of 2D topological field theories*, Integral systems and quantum groups, LNM 1620, Springer, 1996, pp. 120–348.

[E-K]   L. Ernström, G. Kennedy, *Recursive formulas for the characteristic numbers of rational plane curves*, alg-geom/9604019 (1996).

[E-S]   G. Ellingsrud, A. Strømme, *Bott's formula and enumerative geometry*, J. Amer. Math. Soc. **9** (1996), 175–193.

[Faber]   C. Faber, *A conjectural description of the tautological ring of the moduli space of curves*, preprint (1996).

[Faber2]   C. Faber, *A non-vanishing result for the tautological ring of $\mathcal{M}_g$*, preprint (1995).

[F-O]   K. Fukaya, K. Ono, *Arnold Conjecture and Gromov-Witten invariant*, Warwick preprint **29** (1996).

[Fulton]   W. Fulton, *Intersection Theory*, Springer Verlag, 1984.

[FultonSC]   W. Fulton, *Enumerative geometry via Quantum Cohomology (An introduction to the work of Kontsevich and Manin)*, Notes handed out at Santa Cruz (1995).

[F-P]   W. Fulton, R. Pandharipande, *Notes on Stable Maps and Quantum cohomology*, alg-geom/9608011 (1996).

[Ginzburg]   V. Ginzburg, *Equivariant cohomology and Kähler geometry*, Funct. Anal. Appl. **21** (1987), 271–283.

[Givental]   A. Givental, *Equivariant Gromov-Witten Invariants*, I.M.R.N. **13** (1996), 613–663.

[G-P]   L. Göttsche, R. Pandharipande, *The quantum cohomology of blow-ups of $\mathbb{P}^2$ and enumerative geometry*, alg-geom/9611012 (1996).

[J-K]   T. Johnsen; S. Kleiman, *Rational curves of degree at most 9 on a general quintic threefold*, Comm. Algebra **24** (1996), 2721–2753.

[Kapranov]   M. Kapranov, *Veronese curves and Grothendieck-Knudsen moduli space $\overline{M}_{0,n}$*, J. of Alg. Geom. **2** (1993), 239–262.

[Keel]   S. Keel, *Intersection theory on moduli spaces of stable n-pointed curves of genus zero*, Trans. of the Am. Math. Soc. **330** (1992), 545–574.

[K-L]   G. Kempf, D. Laksov, *The determinantal formula of Schubert calculus*, Acta math. **132** (1974), 153–162.

[KSV]   T. Kimura, J. Stasheff, A. Voronov, *On operad structures of moduli spaces and string theory*, Comm. Math. Phys. **171** (1995), 1–25.

[Kleiman]   S. Kleiman, *The transversality of a general translate*, Compositio Math. **38** (1974), 287–297.

[Knudsen]   F. Knudsen, *Projectivity of the moduli space of stable curves. II*, Math. Scand. **52** (1983), 1225–1265.

[Kontsevich]   M. Kontsevich, *Enumeration of rational curves via torus action*, The moduli space of curves, (Dijkgraaf, Faber, van der Geer, editors) Birkhäuser, 1995, pp. 335–368.

[Kontsevich2]   M. Kontsevich, *Intersection theory on the moduli space of curves and the matrix Airy function*, Comm. Math. Phys. **147** (1992), 1–23.

[K-M]   M. Kontsevich, Y. Manin, *Gromov-Witten classes, quantum cohomology, and enumerative geometry*, Comm. Math. Phys. **164** (1994), 525–562.

[Kresch]   A. Kresch, *Associativity relations in quantum cohomology*, Preprint (1996).

[Li-Tian]   J. Li, G. Tian, *Virtual moduli cycles and Gromov-Witten invariants of algebraic varieties*, `alg-geom/9602007` (1996).

[Li-Tian2]   J. Li, G. Tian, *Virtual moduli cycles and Gromov-Witten invariants of general symplectic manifolds*, `alg-geom/9608032` (1996).

[Looijenga]   E. Looijenga, *On the tautological ring of $\mathcal{M}_g$*, Invent. Math. **121** (1995), 411–419.

[May]   J. P. May, *Definitions: Operads, Algebras, and Modules*, Operads: Proceedings of Renaissance conference, Cont. Math., vol. 202, 1997.

[May2]   J. P. May, *The geometry of iterated loop spaces*, LNM 271, Springer, 1972.

[Mumford]   D. Mumford, *Towards an enumerative geometry of the moduli space of curves*, Arithmetic and Geometry II, (M. Artin and J. Tate, editors) Birkhäuser, 1983.

[P]   R. Pandharipande, *Intersections of $\mathbb{Q}$-divisors on Kontsevich's moduli space and enumerative geometry*, `alg-geom/9504004`.

[Siebert]   B. Siebert, *Gromov-Witten invariants of general symplectic manifolds*, preprint `dg-ga/9608005` (1996).

[Stasheff]   J. Stasheff, *The pre-history of operads*, Operads: Proceedings of Renaissance conference, Cont. Math., vol. 202, 1997.

[Strømme]   A. Strømme, *On parametrized rational curves in Grassmann varieties*, LNM 1266, 1987, pp. 251–272.

[Thomsen]   J. Thomsen, *Irreducibility of $\overline{M}_{0,n}(G/P,\beta)$*, `alg-geom/9707005` (1997).

[Witten]   E. Witten, *Two-dimensional gravity and intersection theory on moduli space*, Surveys in Diff. Geom. **1** (1991), 243–310.

—P. Aluffi
Mathematics Department
Florida State University
Tallahassee, FL 32306, USA
aluffi@math.fsu.edu

—P. Belorousski
Department of Mathematics
University of Chicago
Chicago, IL 60637, USA
pavel@math.uchicago.edu

—C. Faber
Department of Mathematics
Oklahoma State University
Stillwater, OK 74078, USA
cffaber@math.okstate.edu

—W. Fulton
Department of Mathematics
University of Chicago
Chicago, IL 60637, USA
fulton@math.uchicago.edu

—T. Graber
Department of Mathematics
University of Chicago
Chicago, IL 60637, USA
graber@math.uchicago.edu

—S.L. Kleiman
Dept of Math, Room 2-278, MIT
77 Massachusetts Ave
Cambridge, MA 02139, USA
kleiman@math.mit.edu

—R. Pandharipande
Department of Mathematics
University of Chicago
Chicago, IL 60637, USA
rahul@math.uchicago.edu

—K. Ranestad
Department of Mathematics
University of Oslo
0316 Oslo, Norway
ranestad@math.uio.no

—J. Thomsen
Matematisk Institut
Aarhus Universitet
DK-8000 Aarhus C, Denmark
funch@mi.aau.dk

—V. Batyrev
Mathematisches Institut
Universität Tübingen
72076 Tübingen, Germany
batyrev@bastau.mathematik.uni-tuebingen.de

—I. Ciocan-Fontanine
Department of Mathematics
Oklahoma State University
Stillwater, OK 74078
ciocan@math.okstate.edu

—B. Fantechi
Dipartimento di Matematica
Università di Trento
38050 Povo, Italy
fantechi@science.unitn.it

—L. Göttsche
I.C.T.P.
P.O. Box 586
34100 Trieste, Italy
gottsche@ictp.trieste.it

—B. Kim
Department of Mathematics
University of California
Davis, CA 95616, USA
bumsig@math.ucdavis.edu

—A. Kresch
Department of Mathematics
University of Chicago
Chicago, IL 60637, USA
kresch@math.uchicago.edu

—Z. Ran
Mathematics Department
University of California
Riverside, CA 92521, USA
ziv@math.ucr.edu

—E. Rødland
Department of Mathematics
University of Oslo
0316 Oslo, Norway
einara@math.uio.no

—E. Tjøtta
Department of Mathematics
University of Bergen
5008 Bergen, Norway
Erik.Tjotta@mi.uib.no

Elenco dei volumi della collana
"Appunti"
pubblicati dall'Anno Accademico 1994/95

GIUSEPPE BERTIN (a cura di), *Seminario di Astrofisica*, 1995.

GIUSEPPE DA PRATO, *Introduction to Differential Stochastic Equations*, 1995.

EDOARDO VESENTINI, *Introduction to continuous semigroups*, 1996.

LUIGI AMBROSIO, *Corso introduttivo alla Teoria Geometrica della Misura ed alle Superfici Minime*, 1997.

CARLO PETRONIO, *A Theorem of Eliashberg and Thurston on Foliations and Contact Structures*, 1997.

MARIO TOSI, *Introduction to Statistical Mechanics and Thermodynamics*, 1997.

MARIO TOSI, *Introduction to the Theory of Many-Body Systems*, 1997.

PAOLO ALUFFI (a cura di), *Quantum cohomology at the Mittag-Leffler Institute*, 1997.

"CompoMat" Loc. Braccone, 02040 Configni (RI), Italy
Finito di stampare nell'aprile 1998